201 Current Topics in Microbiology and Immunology

Springer
Berlin
Heidelberg
New York
Barcelona
Budapest
Hong Kong
London
Milan
Paris
Tokyo

An Antigen Depository of the Immune System: Follicular Dendritic Cells

Edited by Marie H. Kosco-Vilbois

With 39 Figures

Springer

Marie H. Kosco-Vilbois
Glaxo Institute for Immunology
14, Chemin des Aulx
1228 Plan-les-Ouates
Switzerland

Cover illustration: The cover illustration which combines a background of scanning electron micrographs with two contrasting cartoon overlays, is designed to emphasize the changing morphology of the FDC from filiform to beaded dendrities during its functional relationship with B cells of the developing germinal center. The yellow overlay depicts the mystery of the FDC (blue)-antigen (red)-iccosome (beads)-B cell (maroon) microenvironment. In this unique setting, the FDC transfers the retained antigen via iccosomes to germinal center B cells. The blue overlay shows the B cell endocytosing the iccosomes for antigen-processing. Via the Golgi apparatus and transport vesicles, the processed iccosomal antigens are then re-expressed at the B cell's surface for presentation to T cells. (Professor Andras K. Szakal, Department of Anatomy, Medical College of Virginia, Virginia Commonwealth University, Richmond, Virginia)

Cover design: Künkel+Lopka, Ilvesheim

ISSN 0070-217X
ISBN-13:978-3-642-79605-0 e-ISBN-13:978-3-642-79603-6
DOI: 10.1007/978-3-642-79603-6

Library of Congress Catalog Card Number 15-12910
Softcover reprint of the hardcover 1st edition 1995

Typesetting: Thomson Press (India) Ltd, Madras
SPIN: 10490011 27/3020/SPS – 5 4 3 2 1 0 – Printed on acid-free paper

Preface

Follicular dendritic cells (FDC) are unique among cells of the immune system. While their morphological characteristics resulted in their inclusion as a 'dendritic cell type', they differ quite significantly from the other members of the dendritic cell family. In contrast to T-cell-associated dendritic cells or the Langerhans cells found in the skin, FDC reside in highly organized B cell follicles within secondary lymphoid tissues. This site of residence provided a nomenclature committee in 1982 with the second descriptive factor for the derivation of their name. The cardinal feature of FDC is to trap and retain antigen on the surface of their dendritic processes for extended amounts of time and it is this feature that provides the conceptual component for the title of this book.

In response to an antigenic challenge, primary B cell follicles undergo dynamic events, giving rise to germinal centers which are associated with activation, expansion, and differentiation processes of B cells. The interactions of B cells with FDC and T cells in the germinal centers are essential for generating the complete repertoire of antibody isotypes obtained during an antibody response. In addition, stimuli either initiated or maintained during the germinal center reponse leads to production of high affinity antibodies through the processes of somatic mutation and clonal selection. In this context, FDC act as a pivotal source of antigen. They accumulate foreign proteins (e.g., viruses and bacteria) in an unprocessed form on their plasma membranes. They then alter their morphology, forming iccosomes, and deliver the immunogen to specific B cells which subsequently internalize, process, and present the material to elicit help from the surrounding T cells. As the response continues, FDC appear to provide soluble and cell-surface-associated stimuli that aid in directing the reaction. Finally, as the germinal center response subsides, FDC undergo morphological changes that cause the immune complexes to become buried within labyrinthian patterns of membrane formed by the braiding of their dendritic processes. This capacity to preserve unprocessed antigen within the body has been linked to the long-term maintenance of circulating high-affinity antibody and B memory cells.

In addition to the benefits associated with this feature of long-term antigen retention, viral trapping is especially significant in the pathogenesis associated with the acquired immunodeficiency syndrome (AIDS). It has now gained widespread acceptance that the FDC-retained human immunodeficiency virus (HIV) provides an important reservoir for this infectious agent. FDC are also involved in other pathological conditions such as certain neoplasias, and ectopic germinal centers are associated with autoimmune diseases. Thus continued accumulation of information for the role of FDC during maintenance of a healthy individual as well as their contribution to disease states will provide insight into means of affecting the humoral and cellular immune response.

The contributions within this book represent reviews of recent work carried on by many of the main laboratories worldwide pursuing questions concerning FDC. The influence of FDC on B cell activation, survival and differentiation are addressed using several different model systems. While the full extent of the molecules involved in these interactions are yet to be revealed, the latest experimental data are presented. In addition to the discussion of their accessory activities, the highly controversial issue for the origin of FDC is also included. While the reader will find a wealth of important information contained in these chapters, hopefully, one will also begin to realize the multitude of questions that remain to be both addressed and answered concerning the full involvement of FDC in immune responses.

Geneva MARIE H. KOSCO-VILBOIS

List of Contents

List of Contributors

(Their addresses can be found at the beginning of their respective chapters.)

A Theory of Follicular Dendritic Cell Origin

A.K. SZAKAL[1], Z.F. KAPASI[2], S.T. HALEY[1], and J.G. TEW[2]

1 Introduction

The currently favored follicular dendritic cell (FDC) derivation theory holds that FDC are of local origin, probably developing from fibroblastic or primitive reticular cells (KAMPERDIJK et al. 1978; GROSCURTH 1980; HEUSERMANN et al. 1980a,b; IMAI et al. 1983; DIJKSTRA et al. 1984; HUMPHREY et al. 1984) or from mesenchymal cells (pericytes around capillaries) found locally in follicles (RADEMAKERS et al. 1988). In early studies, FDC were identified as a reticular cell type (MITCHELL and ABBOT 1965; SZAKAL and HANNA 1968), primarily on account of the similarities between reticular cell and FDC nuclei. Both cell types possess a highly euchromatic nucleus with some marginated chromatin. However, the nuclei of FDC tend to be more pleomorphic and to exhibit a greater variety of nuclear shapes, lobations, and numbers of nuclei. Cells identified by electron microscopic antigen localizations as FDC were occasionally also shown to be closely associated with reticular fibers (MITCHELL and ABBOT 1965; DIJKSTRA et al. 1984). HUMPHREY et al. (1984)

[1]Department of Anatomy, Division of Immunobiology, Medical College of Virginia, Virginia Commonwealth University, P.O. Box 980709, Richmond, VA 23298, USA
[2]Department of Microbiology/Immunology, Medical College of Virginia, Virginia Commonwealth University, P.O. Box 980678, Richmond, VA 23298, USA

reported that FDC in bone marrow chimeras were of the host phenotype, which further strengthened the belief that FDC are of local derivation. Our report on antigen transport in the lymph node (SZAKAL et al. 1983) suggested that antigen-transporting cells (ATC) were pre-FDC, originating outside of the lymph node. We thus began to question the FDC local origin theory. In addition to the descriptive antigen transport data, we obtained supporting data from experiments conducted with bone marrow chimeras (KAPASI et al. 1993). In this chapter, we will give an account of our studies supporting the contention that FDC are derived from tissues outside of the lymph node. A model summarizing our current thinking is presented in the final section of this paper.

2 Does Antigen Transport by Antigen-Transporting Cells Support the Idea of an Extranodal Derivation of Follicular Dendritic Cells?

Antigen transport to lymph node follicles was first reported in 1983 (SZAKAL et al. 1983) and subsequently described in detail in several reviews (TEW et al. 1984; SZAKAL et al. 1988).

Antigen transport, a cell-mediated mechanism, involves the migration of ATC to follicles during the first 24–36 h after antigenic challenge (SZAKAL et al. 1983). Antigen transport studied in C3H and C57BL/6 mice can be recognized by light microscopy by the path of immune complex-coated ATC as early as 1 min after antigen injection of passively immunized mice. Each path fans out from a narrow section of the subcapsular sinus toward a follicle in the cortex and appears in tissue sections as a triangular peroxidase-positive area when using horseradish peroxidase (HRP) as the antigen or when labeling these sites with monoclonal antibodies (SZAKAL et al. 1983). Control mice lacking the antigen specific antibodies do not show any cell-mediated antigen transport after antigen injection. Consequently, this mechanism is antibody dependent and primarily functional during the late phases of the primary and the secondary antibody responses. It should be pointed out that the immune complexes transported in this model were formed in vivo under physiological conditions. In contrast, preformed immune complexes used for the demonstration of this antigen transport mechanism by others were unsuccessful (KAMPERDIJK et al. 1987; HEINEN et al. 1986). To understand this discrepancy, we studied antigen transport after injection of varying doses of antigen and passively administered specific antibody. We observed (Z.F. KAPASI et al., unpublished) that using a particular batch of anti-HRP, when 10–15 μg antigen were injected in the footpads, as opposed to the routinely administered 1-5 μg HRP in the footpads, the antigen transport site tended to be overwhelmed by antigen draining freely through intercellular spaces of the cortex toward the medulla. This made the recognition of the antigen transport sites impossible. By increasing the dose of the passively administered antibody or reducing the amount of the challenge antigen, fewer immune complexes were transported, which also made the recognition of antigen transport sites difficult. These observations suggest that the antigen to antibody ratio of the immune complexes formed in vivo may also relate to the difficulty or ease of detection of this cell-mediated antigen transport.

During antigen transport, the first cells observed with immune complexes on their surfaces are found in the afferent lymph of the subcapsular sinus (SZAKAL et al. 1983). These cells have a monocyte-like morphology, an indented relatively euchromatic nucleus, and small, peroxidase-positive cytoplasmic granules (primary lysosomes) asociated with the Golgi apparatus (SZAKAL et al. 1983). On these cells, surface area is increased by numerous veils which bear a relatively large immune complex load. These nonphagocytic cells are quite distinct from the typical phagosome-rich sinusoidal macrophages that phagocytize the majority of immune complexes (SZAKAL et al. 1983). There is always a distinct spacial separation between these veiled and dendritic ATC and the antigen-laden macrophages in the subcapsular sinus. This cell-mediated antigen transport pathway is focused on specific anatomical sites of the subcapsular sinus floor typically located over follicles.

These ATC penetrate the subcapsular sinus floor through pores, and this passage through a pore can be convincingly followed by electron microscopy (SZAKAL et al. 1983). As these cells move below the sinus floor, their nuclei increase in size and number of lobes. At this time, ATC begin to develop long dendritic processes, which interdigitate with processes of other ATC in the antigen transport pathway. Thus, these ATC form a chain—a reticulum—which expands toward the nearby follicle in the cortex with the arrival of new ATC. ATC become increasingly more dendritic along this antigen transport chain, and as the cells enter the follicles they take on the morphology of mature FDC. By 24 h after antigen injection, the FDC antigen-retaining reticulum is established at most sites through this mechanism.

In our original paper on antigen transport, we proposed three hypothetical mechanisms of antigen transport. According to one mechanism, ATC may be cells strictly concerned with the transport and transfer of antigen to FDC. According to the second hypothetical mechanism, ATC may be pre-FDC that mature to become FDC (as described above) and would not involve a transfer of the transported antigen. A thrid alternative may be a combination of these two mechanisms. We reasoned that if ATC were pre-FDC, then their antigenic phenotype should be either very similar or identical to that of mature FDC. To determine this, we phenotyped ATC in situ between 1 and 15 min after antigen injection in parallel with mature FDC at 3 days after antigen injection. The results showed that ATC and FDC were of the same phenotype (Table 1). From this phenotyping and the observed maturation of ATC in the transport chain, as indicated by morphology, we concluded that ATC are pre-FDC. Furthermore, since the first monocyte-like, veiled ATC were seen in the afferent lymph of the subcapsular sinus, we proposed that FDC are derived from monocyte-like, veiled cells originating outside of the lymph node.

The recognition of this cell-mediated, antibody-dependent mechanism of antigen transport to lymph node follicles prompted some pertinent questions.

Table 1. Antigen transport cell (ATC) versus follicular dendritic cell (FDC) phenotype in situ

Monoclonal antibodies	Rat isotype	Anti-mouse reactivity	ATC 1–15 min after Ag injection	FDC 3 days after Ag injection
FDC-M1	IgG	FDC and TBM	+	+
MK-1	IgG2a	ICAM-1	+	+
2.4G2	IgG1	Low-affinity FcγRII	+	+
8C12	IgG	CR1	+	+
F4/80	IgG2b	Mo, Mϕ, and LC	+	+
Mac-2	IgG2a	Mϕ subset	–	–
MOMA-2	IgG2b	Mo, Mϕ, and LDC ±	+	+
MIDC-8	IgG2a	DC cytoplasmic, VC, LC	+	+
NLDC-145	IgG2a	Cytoplasmic and membrane IDC, VC, LC	+	+
Anti-CD45 M1/9.3	IgG	Ly-5 (T200) on T and B cells and Mϕ	+	+

Ig, immunoglobulin; ICAM, intercellular adhesion molecule; Ag, antigen; FcγR, Fcγ receptor; CR, complement receptor; Mo, monocyte; Mϕ, macrophage; LC, Langerhans cell; TBM tingible body macrophage; DC, dendritic cell; LDC, lymphoid dendritic cell; VC, veiled cell; IDC, interdigitating cell; Ly, lymphoid.

2.1 Are There Enough Antigen-Transporting Cells to Transport the Antigen?

At 24 h, the fully developed antigen-retaining reticulum is estimated to contain about 2500 FDC. This estimate is based on size measurements of single FDC and the size of the entire FDC antigen-retaining reticulum. Electron microscopy suggests that the entire antigen transport path at 1 min after antigen injection consists of approximately 500 ATC. Thus, at least 500 cells per follicle and, in the case of an entire popliteal lymph node (typically eight follicles), about 4000 transport cells would be available in the vicinity of the injection site during the first minute of transport. More ATC are assumed to be dispersed in the connective tissue of the skin and more can come from the circulating population of precursors in the blood. The total number of ATC required for the full development of FDC networks (an average of eight) in the popliteal lymph node is approximately 20×10^3 ATC, a relatively small number, which it is reasonable to think would be available.

2.2 Can the Amount of Antigen Transported by Antigen-Transporting Cells Account for the Amount of Antigen Retained by Follicular Dendritic Cell Networks?

As antigen transport progresses, each new incoming ATC interdigitates through its developing antigen-bearing processes with the preceding ATC. It follows then that the amount of antigen in the FDC network is equal to that transported by the total number of ATC that make up a particular antigen-retaining FDC network. In contrast to the putative FDC precursors observed by others to be carrying only a few, preformed, gold bead-conjugated immune complexes (Heinen et al. 1986)

ATC are heavily coated with HRP–anti-HRP complexes (SZAKAL et al. 1983). Clearly, these migrating pre-FDC are covered with enough immune complexes (formed under physiological conditions in vivo) to account for the antigen retained in the FDC network (SZAKAL et al. 1983).

Numerous attempts in different laboratories have failed to demonstrate any proliferative activity of FDC residing in the FDC- network. However, the stability of immune complexes retained in the FDC- network for long periods of time (months to years) is well known (TEW et al. 1990). It was also shown that the immune complexes retained on the FDC could not be chased out with antigens of different specificities (TEW and MAMDEL 1978). This apparently static nature of FDC networks initiated questions such as: What happens when new immune complexes are transported to an existing FDC network? Are the new ATC added to the reticulum? Would that not eventually make the reticulum too large?

We know that, in the FDC network, FDC with filiform dendrites mature to form beaded dendrites to produce iccosomes (SZAKAL et al. 1988, 1990). Iccosomes, the small immune complex-coated spheres, are endocytozed by specific B cells in the follicle which process the iccosomal antigen and present it to T helper cells (KOSCO et al. 1988). Data shows that iccosome formation and dispersion may dissipate part or all of an FDC network (SZAKAL et al. 1990). Consequently, there are periods in which FDC and the associated antigen are conserved or retained and periods in which some or all of the antigen may be used up. For example, after an immune response when there is an excess of circulating antibodies, FDC dendrites are in the "ball-of-yarn" configuration for long-term retention of antigen (SZAKAL et al. 1988, 1989). We reasoned that when a new dose of antigen eliminates this antibody excess, it activates the observed maturation sequences of FDC for iccosome production, B cell endocytosis of iccosomes, antigen processing and presentation to T helper cells, resulting in a new cycle of antibody synthesis (SZAKAL et al. 1989; TEW et al. 1990). On account of the dispersion and endocytosis of iccosomes, this process is again accompained by a partial or complete dissipation of the FDC networks. Therefore, the mechanisms of antigen transport and FDC network dissipation appear to be in balance with each other and will tend to prevent the development of overly large FDC networks.

2.3 What Happens When an Antigen of Different Specificity Is Transported to Follicles Already Occupied by an Immune Complex-bearing Follicular Dendritic Cell Network?

A recent study in our labortary on serial antigen administration (i.e., HRP injection is followed by injection of the same molar quantity of alkaline phosphatase 3 days later in a mouse passively immunized for both antigens) showed a layering out of the two antigens in the same FDC network. A layer of HRP (brown), the antigen injected first, was adjacent to the forming germinal center at 4 days, while the antigen, alkaline phosphatase (blue), injected 3 days later was localized more peripherally and appeared to be stacked on top of the FDC retaining the HRP. The FDC retaining the first antigen (HRP) were not randomly mixed with the FDC retaining the second antigen (alkaline phosphatase; blue). Therefore, if the antigen dose is sufficiently large to induce antigen transport to *all* follicles, along with germinal center dissociation—an antigen dose-dependent phenomenon causing dispersion of preexisting germinal center cells (HANNA et al. 1966; SZAKAL et al. 1990)—a displacement of the preexisting FDC networks may also occur. However, small antigen doses differing in specificity from any already retained antigen may simply add to the preexisting FDC networks rather than replacing them.

To summarize this section, it appears that the cell-mediated antigen transport mechanism can account for the transport of sufficient amount of antigen to follicles. Concomitantly, this antigen transport mechanism also provides pre-FDC (ATC) of an identical phenotype with FDC. The pre-FDC mature in transit from the afferent lymph to the follicles and ultimately form the FDC antigen-retaining reticulum. By working in concert with the mechanisms of iccosome formation and dispersion that result in partial or complete dissipation of FDC networks, a relatively steady state in the size of FDC networks may be achieved. Antigen transport, as it exists in the lymph node, supports the extranodal FDC derivation theory.

3 Do Studies on Severe Combined Immunodeficient Mouse/Bone Marrow Chimeras Support Follicular Dendritic Cell Bone Marrow Origin?

We were prompted to use severe combined immunodeficient (SCID) mice for the construction of chimeras, knowing that FDC were radioresistant to high doses of radiation (i.e., 1600 rad, G.F. BURTON, unpublished; 18.5 Gy, NETTESHEIM and HANNA 1969). This was a critical observation, especially since we had confirmed the results reported by HUMPHREY et al. (1984) that FDC are of the host origin in bone marrow chimeras (M.H. KOSCO and G.F. BURTON, unpublished). Thus, if host FDC cannot be eliminated from the system by radiation, then their presence may have inhibited donor FDC from developing normally. To circumvent this FDC radioresistance problem, we turned to the SCID mouse model. SCID mice lack functional B and T cells (BOSMA et al. 1983). As we have demonstrated, they also lack FDC (KAPASI et al. 1993). The absence of FDC was shown by the lack of antigen localization on FDC in lymph node and splenic follicles and by the lack of labeling with the monoclonal antibody FDC-M1 (KAPASI et al. 1993). We reasoned that the SCID mouse would provide a useful model to study the origin of FDC, since the problem of eliminating radioresistant FDC could be bypassed. However, since reconstitution studies with B and T cells resulted in the development of FDC networks, these studies clearly showed that SCID mice have FDC precursors (KAPASI et al. 1993). In ontogeny, FDC do not appear until about 3 weeks after birth (DIJKSTRA et al. 1982,1984; HOLMES et al. 1984). Therefore, we reasoned that transferring bone marrow or fetal liver cells to newborn SCID mice would be advantageous, and donor pre-FDC would not have to compete with host-origin FDC precursors for location or development into mature FDC (see discussion below).

Since studies from other laboratories have demonstrated that SCID mice accept xenogeneic transplants in addition to F_1(Balb/c x C57BL) bone marrow, we also reconstituted SCID mice with rat bone marrow or rat fetal liver cells. We then evaluated the recipients for rat donor phenotype FDC, 6–8 weeks after rat bone marrow or rat fetal liver cell transfers, using the mouse–anti-rat FDC-specific monoclonal antibody ED5. Recipients of F_1 mouse bone marrow were evaluated 6 months after cell transfer for donor-phenotype FDC using monoclonal antibodies against C57BL/6 class I antigens (KAPASI et al. 1994). According to the results of phenotyping, of seven rat bone marrow-reconstituted SCID mice, three mice clearly showed the presence of $ED5^+$ FDC networks in lymph nodes and spleens. Similarly, one out of five recipients of rat fetal liver also had rat FDC in their lymphoid organs. We also noted that a given follicle tended to have either donor or host FDC predominating, although there were follicles with both cell types. It should be noted that FDC-$M1^+$ FDC were not reactive with ED5, indicating that the monoclonal antibody ED5, as reported (JEURISSEN and DIJKSTRA 1986), does not cross-react with mouse FDC. Similarly, the $ED5^+$ cells did not cross-react with the monoclonal antibody FDC-M1. The presence of host FDC in our recipients is not surprising, considering that SCID mice have FDC precursors. Nevertheless, the presence of FDC networks of donor and donor–host phenotypes were the major findings (KAPASI et al. 1994). The results showed that bone marrow and fetal liver both contained FDC precursors. Chimeras constructed with F_1 bone marrow confirmed these observations. These observations also give credence to the pre-FDC nature of ATC and to the extranodal origin of FDC.

Further confirmation of the presence of FDC precursors in the bone marrow was obtained using the *lacZ* mouse model (SANES et al. 1986). Rosa-26 mice transfected with the *lacZ* gene express the gene product β-galactosidase in all cells. Through the action of this gene product, the fluoresceinated substrate fluorescein-di-β-galactopyranoside (FDG) (MOLECULAR PROBES 1989; ROEDERER et al. 1991) is cleaved and the fluorescein is released into the cytoplasm. As a result, the cell becomes fluorescent and detectable by flow cytometry. For electron microscopic identification of the *lacZ* product, the substrate X-Gal is used (SANES et al. 1986). Using the same protocol for the construction of chimeras as above, newborn (3-day-old) SCID mice received bone marrow cell transfers from Rosa-26 mice. We reasoned that if the SCID mice were repopulated by Rosa-26 bone marrow FDC precursors, then the FDC networks would be identifiable through the presence of the *lacZ* gene product. For flow cytometry control, FDC were isolated according to protocol (Schnizlein et al. 1985) from Balb/c mice. The enriched preparation contained 23% FDC. By FACScan (fluorescence-activated cell sorter) after being reacted with biotinylated FDC-M1 and streptavidin-phycoerythrin, 99% of the FDC-M1$^+$ population was shown to be negative for the *lacZ* product. When a similarly enriched preparation (25%) of FDC from *lacZ* mouse bone marrow-recipient SCID mice were tested, the result showed a ratio of 67% host to 33% donor (*lacZ*$^+$) phenotype FDC for the SCID–Rosa-26 chimeras. These results supported the results of our previous SCID mouse–rat and SCID mouse–F_1 chimeras and allowed us to conclude that FDC precursors *can* come from the bone marrow.

4 Can Circulating Follicular Dendritic Cell Precursors Be Identified?

Since our SCID chimera studies indicate a bone marrow precursor for FDC, it was reasonable to expect that these precursors may be identified in the blood. Using the monoclonal antibody KiM4 specific for human FDC, Parwaresch et al. (1983a,b) identified KiM4-positive cells in the human blood. These cells were large (approximately 16 μm in diameter), veiled mononuclear cells. Using the monoclonal antibody FDC-M1 prepared against mouse FDC, we were also able to identify cells 14–16 μm in diameter in the blood (Szakal et al. 1994). These FDC-M1$^+$ cells, just like FDC, were highly reactive for Fc receptors (FcR) using the monoclonal antibody 2.4G2. These cells were found in very low numbers. Further characterization of these blood-borne precursors is presently in progress. The finding of this potential FDC precursor in the blood made the connection between the bone marrow, a source of the precursors, and the tissues drained by the lymph nodes and provided additional support for the idea of the bone marrow derivation of FDC precursors.

5 What Is the Identity of the Follicular Dendritic Cell Precursor in the Bone Marrow?

The main issue regarding the bone marrow derivation of FDC precursors is whether these precursors are derived from hemopoietic cells or stromal cells. We have approached this problem also with the use of the monoclonal antibody FDC-M1. For light microscopic evaluation, we used immunoperoxidase or immunoalkaline phosphatase techniques. For electron microscopy rat–anti-mouse FDC-M1, biotinylated mouse–anti-rat immunoglobulin (Ig G), and streptavidin-peroxidase were used. Preliminary studies identified stromal cells and megakaryocytes labeled with FDC-M1. Inappropriate isotype controls were negative.

Among the stromal cells, we have tentatively identified, with light and electron microscopy, FDC-M1⁺ macrophage-like cells with prominant phagosomes, cells with multiple, elongated cell processes of a supporting reticular cell type, some of which appeared to contain multilobed nuclei or were multinucleated (SZAKAL et al. 1994). At this point, it is uncertain whether these multilobed cells with elongated processes were unusual forms of megakaryocytes or truly represented a distinct cell type. We feel that the FDC bone marrow precursor has not yet been identified, and a hemopoietic or a stromal origin of the FDC precursor are both possibilities. It should be pointed out, however, that, in phenotyping ATC and FDC, in addition to FDC-M1 and several other markers (see Table 1), we found pre-FDC (ATC) and FDC to be positive for the macrophage marker F4/80. Similarly, the phagocytic cells in the bone marrow reacting with FDC-M1 were also reported to be positive for F4/80 (PENN et al. 1993). We wonder, therefore, whether this F4/80/FDC-M1 positivity is a coincidence or a true indication of the bone marrow precursor of FDC. To obtain a conclusive answer regarding the hemopoietic origin of FDC, we will isolate mouse bone marrow cells, using monoclonal antibodies to the SCA-1 antigen shared by hemopoietic bone marrow precursors (VAN DE RIJN et al. 1989), for reconstitution of SCID mice. The presence of *lacZ*⁺ FDC in SCID recipients would indicate a hemopoietic derivation of FDC.

6 A Working Model Summarizing Our Data and Current Concepts on Follicular Dendritic Cell Origin

According to our working model (Fig. 1), FDC originate from the bone marrow. The specific bone marrow precursor has not yet been identified, although bone marrow cells label with the FDC-reactive monoclonal antibody FDC-M1. The FDC precursor enters the blood from the bone marrow to be delivered to the various connective tissue compartments (e.g., dermis). On reaching these connective tissue compartments, FDC precursors are believed to persist until the appropriate stimulus mobilizes them again. These connective tissue pre-FDC or veiled ATC may derive from the blood-borne, large FDC-M1/2.4G2⁺ precursors in mice and from the circulating KiM4⁺ veiled equivalents of these cells in humans. As suggested by PARWARESCH et al. (1983a,b) and by the morphology of our observed ATC arriving at the subcapsular sinus in the afferent lymph (SZAKAL et al. 1983), these cells may be of a monocytic lineage (TEW et al. 1984). In fact, phenotyping studies of ATC and FDC show the presence of cell surface markers such as intercellular adhesion molecule (ICAM)-1, FcγRII, FcεRII, F4/80, MOMA-2, and at least one of the isoforms of leukocyte common antigens (LCA) on the ATC/FDC group, all of which are typical markers of the myeloid lineage. The pre-FDC found

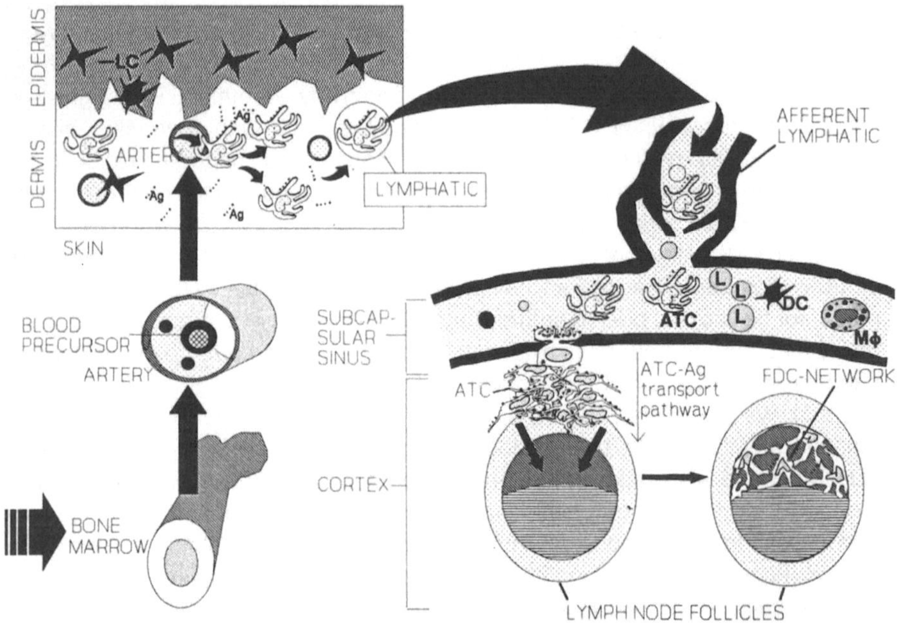

Fig. 1. A working model of the bone marrow derivation of follicular dendritic cells (FDC). The migratory pathway of pre-FDC from the bone marrow (*bottom left*) to the lymph node follicles (*bottom right*). Note the presence of pre-FDC (FDC-M1$^+$ cells) in the circulation and their entry into the connective tissue of the skin, where they bind the antigen in the form of immune complexes and leave the connective tissue via the lymphatics. These pre-FDC (veiled cells) enter the lymph node through afferent lymphatics and arrive in the subcapsular sinus of the lymph node. Via pores in the floor of the subcapsular sinus, pre-FDC antigen (*Ag*)-transporting cells (ATC) enter the cortex and transport the antigen into the follicle, where they mature and form the antigen-retaining reticulum or FDC- network. *L*, lymphocyte; *DC*, dendritic cell; *M*, macrophage; *LC*, Langerhans cell

in the dermis may also be related to precursors of Langerhans cells and dermal dendritic cells (DC). This relationship is suggested by the observation that, in immune mice when the footpad is injected with radiolabeled antigen, the antigen is localized in the epidermal Langerhans cells as well as in dermal mononuclear cells (J.G. Tew, unpublished observation). This relationship is further supported by the MIDC-8 and NLDC-145 reactivity of ATC (pre-FDC) and FDC (see Table 1), two T cell-associated dendritic cell reagents. We attribute this ATC/FDC reactivity in the case of NLDC-145 to the reactivity reported in the periarteriolar lymphocytic sheath (PALS), where normally the FDC network is located. This was identified originally as NLDC-145 reactivity with nonlymphoid cells 4 days after 900 rad irradiation (Kraal et al. 1986). Another reason for this yet unclaimed reactivity with ATC and FDC of these T cell-associated dendritic cell reagents is the use of more sensitive methods than those used originally in the characterization of these antibodies (i.e., frozen sections, reduced fixation: 30 s in acetone; longer incubation times with antibodies, plus use of a biotinylated secondary antibody amplification system; Breel et al. 1987).

Pre-FDC may remain in the dermal connective tissue for a period to mature or perhaps even to replicate. The mobilization stimulus of these connective tissue pre-FDC (ATC) may be a lymphokine, as suggested by the B and T cell requirements of FDC development (KAPASI et al. 1993). This lymphokine release may in turn be stimulated by the presence of specific immune complexes, resulting in the migration of these veiled pre-FDC to the draining lymph nodes, not unlike the migration of Langerhans cells. En route, these precursors bind immune complexes to their surface via Fc and complement receptors and transport these complexes to follicles. When all immune complexes have been phagocytized or bound to ATC and cleared or transported from the connective tissue (e.g., dermis), some of these pre-FDC may still migrate to the follicles without transporting immune complexes. These immature cells may remain in a pre-FDC state. Such dormant ATC would be found, in transit in pores of the subcapsular sinus floor, ready to trap new immune complexes. Other ATC may become attached to reticular fibers or even local follicular capillaries until sufficient immune complexes are bound by them to induce their maturation to FDC. Thus, in situ development of FDC from pre-FDC may also be detected.

We believe the radioresistance of FDC to be a significant problem in radiation chimeras constructed in past studies of FDC origin (HUMPHREY et al. 1984). This may be avoided by the use of SCID mouse recipients which lack mature FDC. However, competition from FDC precursors must also be considered in view of the possible dispersed locations of intermediate forms of pre-FDC. We utilized this idea of pre-FDC dispersion in the construction of SCID–rat bone marrow (or fetal liver) chimeras. Previous reports (DIJKSTRA et al. 1982, 1984; HOLMES et al. 1984) showed that FDC or their precursors are first found in lymphoid tissues at about 3 weeks after birth. For donor FDC precursors to avoid competition from host FDC and their precursors, we selected newborn SCID mice as recipients. We reasoned that this would prevent inhibition of donor pre-FDC migration to lymphoid organs by host FDC precursors found dispersed in the various tissues. We also believe that these dispersed precursors play an important role in antigen transport and repopulation of FDC networks in recipients, whether these are SCID mice or radiation chimeras. For example, even if FDC precursors in the bone marrow are less radioresistant than mature FDC and their intermediate dispersed precursors, the pool of intermediate type pre-FDC could be large enough to supply precursors for months of antigenic stimulation, resulting in the development of new FDC networks.

Our future goals in testing and studying this working model of FDC origin will include the further characterization of FDC-M1/2.4G2$^+$ circulating large mononuclear cells. In addition, pre-FDC population characteristics that must exist in the pathway between the circulating form of the precursor and those which have been characterized as ATC in the afferent lymph are of interest. With regard to the bone marrow, we will characterize the recently found FDC-M1-reactive cells and will isolate hemopoietic stem cells for construction of SCID chimeras to determine whether these stem cells are of hemopoietic or stromal origin.

References

Bosma GC, Custer RP, Bosma MJ (1983) A severe combined immunodeficiency mutation in the mouse. Nature 301: 527–530

Breel M, Mebius R, Kraal G (1987) Dendritic cells of the mouse recognised by two monoclonal antibodies. Eur J Immunol 17: 1555–1559

Dijkstra CD, Van Tilburg NJ, Dopp EA (1982) Ontogenetic aspects of immune complex trapping in rat spleen and popliteal lymph nodes. Cell Tissue Res 223: 545–552

Dijkstra CD, Kamperdijk EWA, Dopp EA (1984) The ontogenetic development of the follicular dendritic cell. An ultrastructural study by means of intravenously injected horseradish peroxidase (HRP)-anti-HRP complexes as marker. Cell Tissue Res 236: 203–206

Groscurth P (1980) Non-lymphatic cells in the lymph node cortex of the mouse. II. Postnatal development of the interdigitating cells and the dendritic reticular cells. Pathol Res Pract 169: 235–254

Hanna MG Jr, Congdon CC, Wust CJ (1966) Effect of antigen dose on lymphatic tissue germinal centers. Proc Soc Exp Biol Med 121: 286–290

Heinen E, Braun M, Coulie PG, Van Snick J, Moeremans M, Cormann N, Kinet Denoel C, Simar LJ (1986) Transfer of immune complexes from lymphocytes to follicular dendritic cells. Eur J Immunol 16: 167–172

Heusermann U, Zuborn K, Schroeder L, Sutte H (1980a) The origin of the dendritic reticulum cell. Cell Tissue Res 209: 279–294

Heusermann U, Zurborn KH, Schroeder L, Stutte MJ (1980b) The origin of the dendritic reticulum cell. An experimental enzyme-histochemical and electron microscopic study on the rabbit spleen. Cell Tissue Res 209: 279–294

Holmes KL, Schnizlein CT, Perkins EH, Tew JG (1984) The effect of age on antigen retention in lymphoid follicles and in collagenous tissue of mice. Mech Ageing Dev 25: 243–255

Humphrey JH, Grennan D, Sundaram V (1984) The origin of follicular dendritic cells in the mouse and the mechanism of trapping of immune complexes on them. Eur J Immunol 14: 859–864

Imai Y, Terashima K, Matsuda M, Dobashi M, Maeda K, Kasajima T (1983) Reticulum cell and dendritic reticulum cell—origin and function. Recent Adv RES Res 21: 51–81

Jeurissen S, Dijkstra C (1986) Characteristics and functional aspects of non-lymphoid cells in rat germinal centers, recognized by two monoclonal antibodies ED5 and ED6. Eur J Immunol 16: 562

Kamperdijk EWA, Dijkstra CD, Dopp EA (1987) Transport of immune complexes from the subcapsular sinus into the lymph node follicles of the rat. Immunobiology 174: 395–405

Kamperdijk EWA, Raaymakers EM, de Leeuw JHS, Hoefsmit EChM (1978) Lymph node macrophages and reticulum cells in the immune response. I. The primary response to paratyphoid vaccine. Cell Tissue Res 192: 1–23

Kapasi ZF, Burton GF, Shultz LD, Tew JG, Szakal AK (1993) Induction of functional follicular dendritic cells development in severe combined immunodeficiency mice: influence of B and T cells. J Immunol 150: 2648–2658

Kapasi ZF, Kosco MH, Shultz LD, Tew JG, Szakal AK (1994) Cellular origin of follicular dendritic cells. Proceedings of the 11th international conference on lymphoid tissues and germinal centers in immune reactions. In: Heinen E, Defresne MP, Boniver J (eds) In vivo immunology: regulatory processes during lymphopoiesis and immunopoiesis. Plenum, New York

Kosco MH, Szakal AK, Tew JG (1988) In vivo obtained antigen presented by germinal center B cells to T cells in vitro. J Immunol 140: 354–360

Kraal G, Breel M, Janse EM, Bruin G (1986) Langerhans cells, veiled cells, and interdigitating cells in the mouse recognized by a monoclonal antibody. J Exp Med 163: 981–997

Mitchell J, Abbot A (1965) Ultrastructure of the antigen-retaining reticulum of lymph node follicles as shown by high resolution autoradiography. Nature 208: 500–502

Molecular Probes (1989) FluoReporter lacZ gene fusion detection kit. Cat No. F-1930/F-1931 for detection of lacZ β-D-galactosidase in single cells. Molecular probes, Eugene, OR

Nettesheim P, Hanna MG Jr (1969) Radiosensitivity of the antigen-trapping mechanism and its relation to the suppression of immune response. Adv Exp Med Biol 5: 167–175

Parwaresch MR, Radzun HJ, Feller AC, Peters KP, Hansmann ML (1983a) Peroxidase-positive mononuclear leukocytes as possible precursors of human dendritic reticulum cells. J Immunol 131: 2719–2725

Parwaresch MR, Radzun HJ, Hansmann ML, Peters KP (1983b) Monoclonal antibody Ki-M4 specifically recognizes human dendritic reticulum cells(follicular dendritic cells) and their possible precursors in blood. Blood 62: 585–590

Penn PE, Jiang DZ, Fei RG, Sitnicka E, Wolf NS (1993) Dissecting the hematopoietic microenvironment. IX. Further characterization of murine bone marrow stromal cells. Blood 81: 1205–1213

Rademakers LHPM, de Weger RA, Roholl PJM (1988) Identification of alkaline phosphatase positive cells in human germinal centers as follicular dendritic cells. Adv Exp Med Biol 237: 165–169

Roederer M, Fiering S, Herzenberg LA (1991) FACS-Gal: flow cytometric analysis and sorting of cells expressing reporter gene constructs. Methods 2: 248–260

Sanes JR, Rubenstein JLR, Nicolas JF (1986) Use of a recombinant retrovirus to study post-implantation cell lineage in mouse embryos. EMBO J 5: 3133–3142

Schnizlein CT, Kosco MH, Szakal AK, Tew JG (1985) Follicular dendritic cells in suspension: identification, enrichment, and initial characterization indicating immune complex trapping and lack of adherence and phagocytic activity. J Immunol 134: 1360–1368

Szakal AK, Hanna MG Jr (1968) The ultrastructure of antigen localization and viruslike particles in mouse spleen germinal centers. Exp Mol Pathol 8: 75–89

Szakal AK, Holmes KL, Tew JG (1983) Transport of immune complexes from the subcapsular sinus to lymph node follicles on the surface of nonphagocytic cells, including cells with dendritic morphology. J Immunol 131: 1714–1727

Szakal AK, Kosco MH, Tew JG (1988) A novel in vivo follicular dendritic cell-dependent iccosome-mediated mechanism for delivery of antigen to antigen-processing cells. J Immunol 140: 341–353

Szakal AK, Kosco MH, Tew JG (1989) Microanatomy of lymphoid tissue during the induction and maintenance of humoral immune responses: structure function relationships. In: Paul WE, Fatham CG, Metzger H (eds) Annual reviews of immunology. Annual Reviews, Palo Alto, pp 91-109

Szakal AK, Taylor JK, Smith JP, Kosco MH, Burton GF, Tew JG (1990) Kinetics of germinal center development in lymph nodes of young and aging immune mice. Anat Rec 227: 475–485

Szakal AK, Kapasi ZF, Haley ST, Tew JG (1994) Multiple lines of evidence favoring a bone marrow derivation of follicular dendritic cells (FDC). In: Banchereau J (ed) Proceedings of the 3rd international symposium on dendritic cells. Plenum, New York

Tew JG, Mandel T (1978) The maintenance and regulation of serum antibody levels: evidence indicating a role for antigen retained in lymphoid follicles. J Immunol 120: 1063–1069

Tew JG, Mandel TE, Phipps RP, Szakal AK (1984) Tissue localization and retention of antigen in relation to the immune response. Am J Anat 170: 407–420

Tew JG, Kosco MH, Burton GF, Szakal AK (1990) Follicular dendritic cells as assessory cells. Immunol Rev 117: 185–211

van de Rijn M, Heimfeld S, Spangrude GJ, Weissman IL (1989) Mouse hematopoietic stem-cell antigen Sca-1 is a member of the Ly-6 antigen family. Proc Natl Acad Sci USA 86: 4634–4638

Follicular Dendritic Cells: Origin and Function

E. Heinen, A. Bosseloir, and F. Bouzahzah

1 Introduction

The human immune system is made up of one trillion lymphocytes. Although dispersed throughout the organism, these cells behave as if they belonged to a single organ. To attain this homeostasis, the lymphoid cells home to and are controlled in lymphoid tissues where, in specific microenvironments, they communicate with each other or with accessory cells and proliferate, mature, or die under stringently controlled conditions. The germinal center microenvironments are among those controlling the B cells which, during the T-dependent humoral immune responses, undergo important phases of their life cycles: activation, proliferation, the isotype switch, affinity maturation, deactivation, apoptosis, etc. The follicular dendritic cells (FDC) are major components of the germinal center microenvironments. Here, we examine their origin and their influence on B cells in the light of recent experimental data.

Institute of Human Histology, University of Liège, rue de Pitteurs, 4020 Liège, Belgium

2 Origin

Identifying the precursor cells which give rise to FDC is difficult, due to the fact that FDC live long, seldom divide, and change their morphology and phenotype during the humoral immune responses.

Different approaches have been developed to determine the origin of the FDC, including morphological studies, ontogenic records, enzymology, immunolabeling techniques, culturing, cell grafting, mRNA analysis, and immortalization. Recently, we reviewed the data relating to the origin of the FDC (HEINEN and BOSSELOIR 1994).

Most electron microscope studies point to a fibroblastic origin: FDC are connected to fibroblasts and collagen bundles; when less differentiated, they resemble fibroblasts (HEUSERMANN et al. 1980; HOEFSMIT et al. 1980; DIJKSTRA et al. 1982; IMAI et al. 1983; RADOUX et al. 1985a). Mature FDC differ from fibroblasts in that they bear a complex network of dendrites retaining an electron-dense material and possess desmosomal and adherent junctions. The perinuclear cytoplasm is thin and contains little ergastoplasmic reticulum. FDC are frequently bi- or multinucleate; their nuclei are indented and filled with euchromatin and are delineated by a thin rim of heterochromatin and a lamina densa. The single nucleolus is usually large. Due to their morphological features, FDC have also been called dendritic reticular cells (NOSSAL et al. 1965) or desmodendritic cells (IMAI et al. 1983).

Their ontogenic development has been followed in rodents, mostly during the postnatal period when the cortex and paracortex differentiate in the lymph nodes. For WILLIAMS and NOSSAL (1966), FDC appear before the lymphoid cells, while for others, the lymph follicles develop prior to FDC differentiation (DIJKSTRA et al. 1982; NAMIKAWA et al. 1986).

Enzymology supports the latter conclusion. Observing the histochemical enzyme pattern during transformation of fibroblastic reticulum cells to typical FDC, HEUSERMANN et al. (1980) found it to change: the cells became decreasingly alkaline phosphatase positive, and a positive alpha naphtyl acetate reaction occurred in the differentiated cells. Recently, BOSSELOIR et al. (1994a) observed prolyl 4-hydroxylase activity in the cytoplasm of human tonsillar FDC, both in situ and after isolation. This enzyme is considered a fibroblast marker and catalyzes the formation of 4-hydroxyproline from proteins containing certain amino acid sequences. RADEMAKERS et al. (1989), studying alkaline phosphatase, proposed a parental relationship between FDC and pericytes. This is not surprising, since pericytes have features of mesenchymal cells.

Immunolabeling experiments have disturbed this consensus that FDC derive from mesenchymal cells. Monocytes/macrophages and FDC share several antigens, i.e., CD11b, CD14, CD16, CD31, major histocompatibility complex (MHC) class II (STEIN et al. 1982; GERDES et al. 1983; TSUNODA et al. 1982 PARWARESCH et al. 1983; HEINEN et al. 1984). Hence, a monocytic origin cannot be excluded, especially since PARWARESCH et al. (1983) found cells in the peripheral blood that express an antigen (Ki-M4) also expressed by FDC. Critics have attacked this conclusion, doubting that certain antigens are truly expressed (CD16, MHC class II). The situation has become even more complex with the discovery, on FDC, of antigens expressed by the lymphoid lineage (CD19, CD21, CD3, CD24, CD37, etc.; JOHNSON et al. 1986; PALLESEN and MYRHE-JENSEN 1987) and with the finding that molecules synthesized by other cells can stick to FDC. Many consider DRC1 antigen to be FDC specific, even though in their original paper, NAIEM et al. (1983), report some positivity among the lymphocytes of the corona. BOSSELOIR et al. (1994b) have demonstrated by cytophotometry that all blood and tissular B cells react with anti-DRC1,while T cells and other cell lineages are negative. RUCO et al. (1991) observed endothelial leukocyte adhesion molecule (ELAM) expression by FDC, a finding which points to an endothelial origin. WACKER et al. (1991) also propose this origin because FDC and endothelial cells (but also monocytes) express the Ki-M4 antigen. SCHRIEVER et al. (1991; SCHRIEVER and NADLER 1992), however, exclude an endothelial origin, since FDC synthesize neither fibronectin nor platelet-derived growth factor (PDGF) receptor-α or -β.

Functional studies have brought some clarification, but also confusion. When injected into animals, labeled immune complexes were retained on fully developed FDC, forming a crescent in the light zone, but also on precursor cells with a mesenchymal cell phenotype (HEINEN et al. 1986). In neonatal animals, they were found to attach first to fibroblast-like cells (IMAI et al. 1983; DIJKSTRA et al. 1984). Similar studies on invertebrates showed the existence of fibroblast-like cells which can retain immune complexes (VAN ROOJEN 1980). SZAKAL et al. (1988) described antigen-transporting cells as large, nonphagocytic cells with lobate or irregular euchromatin-containing nuclei and cell processes located in or near the subcapsular sinus or deeper in the cortex. These cells in or near the sinus looked like monocytes, while the cells near or inside the follicles were like FDC. Antigen transport may thus involve the migration of transporting cells with a concomitant maturation into FDC or transfer from one transporting cell to another, sedentary one. These authors thus favor an exogenous origin for immune complex-laden FDC (which would settle inside the follicles). We injected gold-labeled immune complexes either subcutaneously or intravenously and dissected, shortly after injection, the draining lymph nodes and spleen. At the ultrastructural level, we observed macrophages and lymphocytes in the subcapsular sinus with gold particles on their surfaces or inside endocytic vesicles. Between the sinus and the follicles, we found only lymphocytes with gold particles on their plasma membranes, migrating towards the germinal centers. The results obtained on the spleen were similar. In in vitro tests, we showed that immune complex-laden lymphocytes can contact FDC and transfer the complexes to them (HEINEN et al. 1986). Our results thus contradict the hypothesis of Szakal's group. Moreover, we found poorly differentiated FDC with mesenchymal features which had bound small quantities of immune complexes.

Culturing FDC for longer periods of time is another way to probe the origin of FDC, since cells cultured in vitro readily dedifferentiate, reverting to their original morphology. TSUNODA et al. (1990) succeeded in culturing FDC for long periods (over 150 days). Freshly purified FDC appear in round cell clusters containing both FDC and lymphoid cells. They readily attach to the culture substrate, then flatten and spread out. For the first few days, the lymphoid cells harbored by emperipolesis survive, but by day 5 they die out, so that only the FDC persist. The latter appear extremely flattened, with fibroblast-like extensions, and display no proliferative activity for long periods. They synthesize typical fibroblast vimentin filaments, but neither desmin nor cytokeratin. During the first few days of culture, they cease to express the typical FDC phenotype.

Cell grafting in irradiated animals is viewed as a means of indisputably determining the origin of FDC. Working with hybrid mice and allotypic markers, HUMPHREY et al. (1984), refuted a bone marrow origin. IMAZEKI et al. (1992), in an elegant spleen implantation study coupled with H-2 class I immunotyping, concluded that FDC come primarily from a stationary cell population, not from recirculating cells. KAPASI et al. (1994) presented data supporting the view that FDC are of hematopoietic origin. These authors used severe combined immunodeficient (SCID) mice lacking FDC, reconstituting them with rat bone marrow, rat fetal liver cells, or bone marrow cells from F1 donors (BALB16 x C57B1/6). FDC, labeled with anti-ED5 or anti-FDC-M1 monoclonal antibody (mAb), displayed donor class I molecules on their surface. Since KAPASI et al. used bone marrow or fetal cells, we can suspect transfer of undifferentiated stem cells, and thus also of mesenchymal cells, so these experiments do not indubitably prove a bone marrow origin for FDC. Very recently, YOSHIDA et al. (1994) obtained differentiation of FDC in SCID mice by injecting lymphocytes. The latter induced complement receptor expression on cells of the reticular meshwork in the follicle; subsequent immune complex binding via complement receptors appeared to give the signal for Fc receptor synthesis. YOSHIDA et al. (1994) thus showed the mesenchymal origin of FDC and the influence of lymphocytes and immune complexes of FDC differentiation.

Analysis of mRNA from purified FDC by the single cell polymerase chain reaction (PCR) has been published by SCHRIEVER et al. (1989; SCHRIEVER and NADLER 1991, 1992). FDC apparently present a very peculiar RNA pattern with high expression of the C3d receptor (CD21) but no mRNA for CD20, CD45, CD4, fibronectin, or PGDF receptor-α or -β. These authors thus view FDC as belonging to a unique lineage, rather than having a hematopoietic or fibroblastic origin.

Many investigators have attempted to immortalize FDC using all possible approaches. LINDHOUT et al. (1993) exploited CD21 expression to transform FDC with Epstein-Barr virus (EBV). They obtained very large, slowly duplicating cells with a fibroblast-like morphology but displaying a different phenotype (intercellular adhesion molecule-1, ICAM-1; CD40; CD75). These cells were able to bind nonautologous B cells, to perform emperipolesis, and to prevent apoptosis.

Fully differentiated FDC do not divide. Reports of mitosis or of thymidine incorporation into FDC precursor cells are rare (EVERETT et al. 1967; VILLENA et al. 1983). We favor the hypothesis that during ontogeny of primary follicles or during the enlargement of the germinal centers, FDC differentiate from mesenchymal cells (thus also pericytes), which lose the capacity to synthesize matrix elements to begin producing adhesion molecules, receptors, FDC-specific surface antigens, etc. (Fig. 1). As increasing quantities of immune complexes are retained, imported by lymphoid cells or by other means, the surfaces of the FDC enlarge so that plicae and dendrites develop. Perhaps interactions with stimulated B or T cells also contribute to increasing the FDC membrane surface and to triggering the expression of certain molecules. When isolated FDC are incubated in the presence of gold-labeled immune complexes

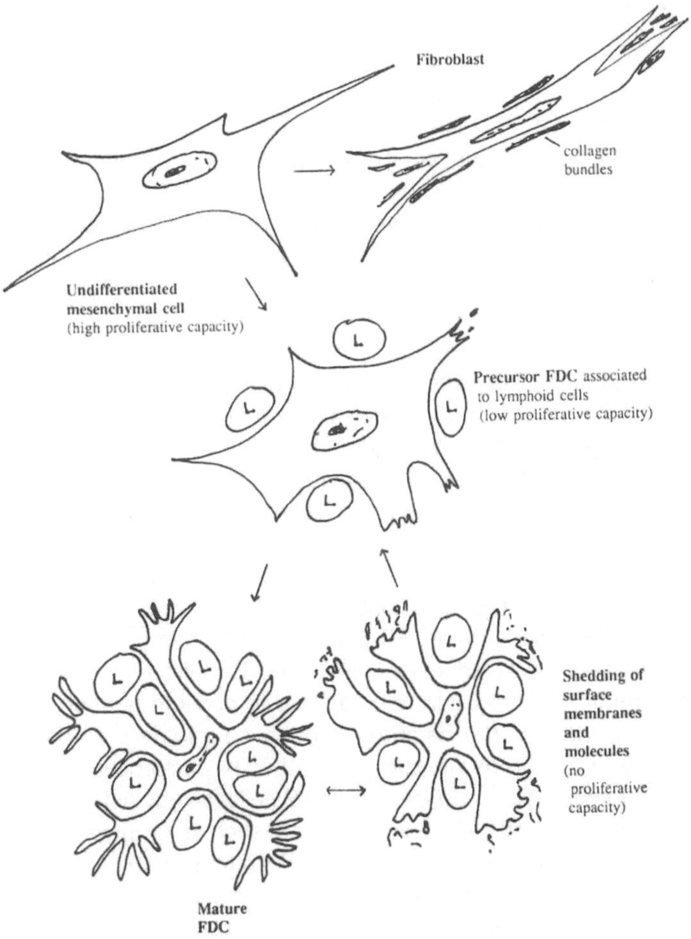

Fig. 1. Hypothetical transformation of mesenchymal cells to fibroblasts or precursors of follicular dendritic cells (*FDC*). Fully developed FDC, in shedding membranes, may regenerate or recover the aspect of precursor cells

and the cells maintained for several hours at 37°C, aggregation of these complexes is observed, followed by their shedding along with the cell membranes (HEINEN et al. 1993). We suggest that fully mature FDC, during and at the end of the germinal center reaction, can release membranes and dendrites with attached material to undergo regeneration or to recover an original, less differentiated morphology and phenotype (Fig. 1).

We have tentatively summarized this proposition in Fig. 1. Mesenchymal cells differentiate into fibroblasts, endothelial cells, and other cells, but also into FDC precursor cells capable of retaining small amounts of immune complexes (IMAI et al. 1983; DIJKSTRA et al. 1984; HEINEN et al. 1986). The induction signals leading to full maturation appear to be given to FDC precursors cells by immune complexes, complement factors, and the associated lymphoid cells.

The retained immune complexes persist for long periods of time but can be renewed (TEW et al. 1979; HEINEN et al. 1983). FDC release immune complexes and cell fragments (HEINEN et al. 1993) which can take the form of beads (iccosomes, SZAKAL et al. 1988; TERASHIMA et al. 1991). In this way, they renew their cell surface or, at the end of the germinal center reaction, recover an undifferentiated, precursor cell aspect. Such intermediate forms between precursors and fully developed FDC as well as degenerative features (dark cells) have been described (RADEMAKERS 1992). As early as 1968, HANNA and SZAKAL noticed that when the germinal centers develop, preexisting FDC increase in size.

FDC thus exhibit various morphological and phenotypic features according to their differentiation stages. This reflects different functional activities and gives rise to different compartments inside the germinal centers HEINEN et al. 1984; IMAI et al. 1986; TERASHIMA et al. 1991; RADEMAKERS 1992; YOSHIDA et al. 1993; HARDIE et al. 1993). In the upper light zone, FDC are fully differentiated with abundant dendrites, huge numbers of C3b and Fc receptors with bound immune complexes. There, FDC strongly express CD21, CD40, CD54, CD71, and CD106 (RICE et al. 1991; CLARK et al. 1992; HARDIE et al. 1993). Mouse FDC express high levels of Fc receptors inside the light zone (RADOUX et al. 1985b; YOSHIDA et al. 1993). In the mantle and dark zones, FDC bear fewer receptors for antibody–antigen complexes and fewer adhesion receptors. B cell survival, activation, proliferation, differentiation, selection, etc., appear to occur in relation to these different microenvironments. The parameters of these microenvironments are influenced not only by the immune complexes, but by reciprocal interactions between FDC and B cells, FDC and T cells, T and B cells, and by external factors: hormones, the nutritional status, exogenous factors such as lipopolysaccharides (LPS) or superantigens (HEINEN et al. 1986, 1990; BRANDTZAEG et al. 1991).

In the following parts of this review, we will consider different FDC functions. Some have been clearly established (effects on B cell proliferation, affinity selection, survival), and others are likely (B cell activation, chemotactism, isotype switch) or hypothetical (inhibition of terminal differentiation, induction of a given T cell phenotype, factor release).

3 Chemotaxis and Adhesion

Lymph follicles are globular structures composed of densely aggregated cells. Chemotactism and intercellular contacts ensure this compactness. FDC attract and bind cells, thus playing a pivotal role in follicle assembly.

Purified germinal center lymphoid cells, labeled and injected intravenously, home to the lymph follicles (OPSTELTEN et al. 1982). To analyze the chemotactic properties of FDC, we placed tonsillar FDC in the lower compartment of Boyden's chambers, with lymph cells deposited in the upper chamber. After 1–2 at 37°C, the lymphoid cells had penetrated further into the filter separating the chambers than when the lower compartment contained only the incubation medium or lymphoid cells. Supernatants taken from cultured FDC affected lymphoid cells similarly. Dilution decreased this effect. The migrating cells were B or T cells. $CD57^+$ T cells were attracted as well as $CD57^-$ T cells (BOUZAHZAH et al., to be published).

In previous papers (LILET-LECLERCQ et al. 1984; HEINEN et al. 1989; LOUIS et al. 1990), we have shown that FDC and lymphoid cells are closely connected and that fluorescent lymphoid cells attach to isolated FDC upon incubation for only half an hour. After 1 and 2 days of culture, very large aggregates are found (HEINEN et al. 1991a). KOSCO et al. (1992) report similar clustering when these cells are cultured together; both B and T cells influence the level of cluster formation. Cell aggregation depends on the presence of divalent ions (Ca^{2+}, Mg^{2+}; SCHRIEVER et al. 1989; LOUIS et al. 1990) and on expression of lymphocyte function-associated antigen (LFA)-1, ICAM-1, vascular cell adhesion molecule (VCAM), (INCAM)-110, and very late antigen (VLA)-4 (FREEDMAN and NADLER 1991; KOOPMAN et al. 1991; KOOPMAN and PALS 1992; PETRASCH et al. 1991b; KOSCO 1991b; KOSCO et al. 1992). Except for those bearing C3 fragments, immune complexes do not appear to influence this adhesion. Anti-DRC1 antibodies slightly reduce B cell adhesion to FDC (LOUIS et al. 1990).

Studies on the transfer of immune complexes to FDC have also revealed migration of lymphoid cells towards the lymph follicles. B and T cells bearing immune complexes bound via C3b and Fc receptors migrate towards FDC, make contact with them, and transfer the complexes to them (BROWN et al. 1973; HEINEN et al. 1984; BRAUN et al. 1987). After irradiation, immune complex retention in the germinal centers can be achieved only when the lymphoid cell populations are reconstituted (KROESE et al. 1986).

Negative chemotactism, i.e., intercellular repulsion or inhibition of migration, may also exist at the level of the germinal centers, since many cell types ($CD8^+$ T cells, granulocytes, mastocytes) are absent or infrequent and the germinal centers are mono- or oligoclonal (KROESE et al. 1987). Besides positive or negative attraction and affinity selection, other mechanisms may control the cells composing the germinal centers, e.g., tingible body macrophages may phagocytose certain cell types, activated B cells may exert dominance, T cells may change their phenotype, immigrated cells may die.

4 Antigen Presentation

"Antigen presentation" is the term that designates the dynamic events preceding the recognition of oligopeptides inserted into MHC molecules by T cell receptors, a process leading either to activation or to tolerization of T cells. This concept is thus restricted to cellular immune responses. Theoretically, for B cells no presentation is required, since B cells recognize native antigen. Inside the germinal centers, however, a peculiar form of antigen presentation to B cells appears to occur: for long periods of time, FDC retain huge amounts of antigen–antibody complexes, bearing complement factors or not. This creates a special context in which germinal center B cells encounter antigen, a context not found elsewhere. These antigen-antibody complexes are highly immunogenic (PHIPPS et al. 1984) and may exert their influence in different ways: the presence of antibody may mask antigenic determinants (TEW et al. 1980), Fc portions may give inhibitory signals (SINCLAIR 1969; MUTA et al. 1994), and the immunoglobulin (Ig) isotypes may exert yet unknown biological actions. Complement split products induce opposite reactions (ERDEI et al. 1991). For example, C3b or C3d (g) inhibits cell growth when soluble, but stimulates it when immobilized. Since inhibitors of the complement cascade are present in the germinal centers (IMAI et al. 1986; LAMPERT et al. 1993), the B cells undego no deleterious action.

When the antigen is densely arranged and ordered, it is stimulatory and tolerization is avoided (BACHMAN et al. 1993). FDC can thus be viewed as effecting a peculiar form of antigen presentation to B cells; this is especially important during affinity selection (see below).

Ultrastructural observations based on the use of gold-labeled antigen or immune complexes have revealed that lymphoid cells are not in contact with all antigenic material, since much of it is inserted in deep FDC invaginations (HEINEN et al. 1983). FDC are thus reservoirs in which antigen is stored for long periods of time (TEW et al. 1979) in the light zone of germinal centers (KLAUS et al. 1980) populated essentially by centrocytes. Since most of these centrocytes become deactivated and transform into memory cells, a contradiction exists between antigen contact and B cell behaviour. The hypothesis of TEW et al. (1980) may provide an explanation: these authors suggest that antibody in excess may mask antigen determinants, thus preventing B cell activation. Otherwise, we must consider the possibility that contact with antigen in a given microenvironment may drive B cells towards deactivation. According to CELADA (1971), for instance, an excess of antibodies may be inhibitory. SZAKAL et al. (1989) have proposed the elegant iccosome (immune complex-coated beads) hypothesis, according to which FDC dendrites fragment into beads covered with immune complexes. These iccosomes would be taken up and processed by B cells for presentation to T cells. We do not agree with this hypothesis for several reasons (CEBRA et al. 1991). First, formation of beads is not specific to FDC; they are observed in vitro with various cell types when dendrites or filopods are formed by retraction when cell extensions degenerate. Second, iccosomes are not frequently encountered in vivo; we think they are part of the above-mentioned cell regeneration process in which differentiated FDC release membranes and receptors to renew their surfaces or to revert to an undifferentiated state. Third, B cells can pick up antigen and process it, but they have so far not been shown to present antigen to germinal center T cells. We propose that germinal center B cells take up antigen retained by FDC, process it, and present it to T Cells located in the T-dependent zones annexed to the follicles. Such antigen uptake and migration before presentation in another tissular compartment is possible and has been demonstrated for dendritic cells: sessile intraepithelial Langerhans cells can mobilize and transform first into veiled cells, then into interdigitating cells. The very attractive iccosome hypothesis, adopted by many immunologists, should thus be considered with caution.

Antigen–antibody complexes retained by FDC serve to periodically restimulate B and perhaps T cells. They may also play a part in the Ig isotype switch and affinity maturation. Such complexes can become highly immunogenic as a result of their density, ordering, and association with complement split products, but interactions with Fc receptors can lead to deactivation (Muta et al. 1994). Consequently, FDC-mediated antigen presentation may be stimulating or suppressive, according to the Ig isotype, antibody concentration, and microenvironment.

5 B Cell Activation

Immune complex retention of FDC generally occurs several days after antigenic stimulation and parallels the transformation of FDC precursors to fully developed FDC. Whether FDC intervene during initial events of virgin B cell activation is thus questionable, since these events take place, according to most authors, in the T-dependent zone (see Gray et al. 1986; Gray and Skarvall 1988; Gray 1989; Heinen et al. 1990). B cell activation takes several hours, required for the sequential regulation of various genes, protein synthesis, and other events (Ashman 1984, 1990). The final activation steps at least, i.e., those leading to cell proliferation, might possibly happen in contact with FDC after migration to the germinal centers. Due to the high density of the immune complexes and to the presence of complement factors, optimal stimulation of B cells could occur. On this basis, Van Noessel et al. (1993) have proposed a model for dual antigen recognition, in which aggregates composed of antibody–antigen complexes and C3dg induce multimerization on the cell surface by formation of bridges between membrane Ig (mIg) and the CD19–TAPA-1–CD21 complexes. This multimerization allows cross-phosphorylation and subsequent activation of cytoplasmic enzymes (src-like protein tyrosine kinase, PTK, lyn, and PTK72). Activated forms of these phosphotransferases may be directly or indirectly responsible for connecting the antigen–receptor complex to second messenger molecules via phosphorylation of tyrosine residues present in the cytoplasmic regions of Igα, Igβ, and CD19. FDC abundantly express various adhesion molecules (β_1, β_2, β_3-integrins, VCAM-1, ICAM-1 and -2, CD40; CLARK and LANE 1991; KOOPMAN and PALS 1992) on their extended cell surface. Ultrastructural studies have revealed intimate contact between FDC and lymphoid cells. One might speculate in two opposite directions. First, receptor cross-linkings could improve B cell survival (CORMANN et al. 1986a; LIU et al. 1989), enabling such B cells to proliferate better than unlinked B cells. Alternatively, when these receptor cross-linkings are numerous and extended over the whole cell surface, the B cells could be in a "frozen state" in which activation and proliferation are slowed down. The first situation could arise in the dark zone, where FDC send few cytoplasmic extensions between B cells. The second could occur in the light zone, where, by emperipolesis, the lymphoid cells (centrocytes) are surrounded by FDC extensions and are clearly less activated than centroblasts. Thus, according to their differentiation state, dictated by the local germinal center microenvironment, FDC might thus influence B cells in two opposite directions. Furthermore, the phenotype of FDC varies according to their location; in the light zone, for instance, the density of retained immune complexes is high. One should bear in mind that contact with FDC is not sufficient for activation or deactivation; other signals are necessary. RODRIGUEZ et al. (1992) have revealed that germinal center B cells inhibit mitogen-induced proliferation of mantle zone B cells. Thus, yet other parameters play a part in the activation or deactivation of follicular cells.

Studying the activating capacity of FDC, KOSCO-VILBOIS et al. (1993) recently showed that B cells maintained in contact with FDC express more B7/BB1 (CD80) surface antigen and can better present antigen to T cells than B cells cultured alone. This interesting observation is somewhat puzzling. First, in primary follicles, FDC are intimately associated with resting B cells, which, in contrast to the above finding, remain in a quiescent state. Most mantle zone B cells appear as long-lived, IgM^+ IgD^+, and anergic (thus tolerized) cells (BEREK and ZIEGNER 1993). Second, T cells inside the germinal centers are not in an activated state (see below). Thus, antigen presentation as proposed by KOSCO-VILBOIS et al. 1993 (see also KOSCO et al. 1988; KOSCO 1991a, b; GRAY et al. 1991) does not occur in the germinal centers but outside, once the B cells have emigrated. Lymph follicles are always associated with T-dependent zones (HEINEN et al. 1990), so the antigen retained by FDC can serve not only to activate B cells but, when borne by B cells having had contact with FDC, to stimulate T cells. Interactions between T-dependent areas and follicles are not unidirectional (stimulation of virgin B cells before they enter the germinal centers), but bidirectional, since emigrating B cells export and present oligopeptides, thereby maintaining T memory cell clones. Third, interaction via MHC class II molecules can effect other signals. The MHC molecules also help transmit signals inside the B cells; for example, binding of anti-MHC II antibodies to B cells can block their multiplication and activation by increasing the intracellular cyclic adenosine monophosphate (cAMP) concentration (NEWELL et al. 1993).

In conclusion, FDC by themselves do not appear to stimulate virgin IgM^+ IgD^+ B cells, but rather sustain the activation of antigen-specific B cells preactivated in the T-dependent area. Deactivation by FDC in the light zone can be suspected, since there centrocytes evolve into small B memory cells.

6 B Cell Proliferation

The link between B cell proliferation and FDC has long been known, since both are observed inside the germinal centers. CORMANN et al. (1986b) demonstrated this link by coculturing isolated FDC and lymphoid cells subjected to different mitogenic challenges. Several authors have confirmed these results (TSUNODA et al. 1989; KOSCO et al. 1988; PETRASCH et al. 1991a). Proliferating cells are associated with FDC: after isolation under mild enzyme treatment, FDC preserve their contacts with lymphoid cells, forming clusters (LILET-LECLERCQ et al. 1984; SCHMITZ et al. 1993); some of these lymphoid cells incorporate tritiated thymidine (CORMANN et al. 1989; KOSCO et al. 1992).

B cell proliferation occurs in the dark and, to a lesser extent, light zone. In the latter, FDC exhibit a peculiar phenotype (CD21, CD23, CD54, calbinding, 5-nucleotidase) and bear large amounts of immune complexes. According to their differentiation state, FDC can thus create a microenvironment that is favorable or not to B cell proliferation or differentiation into memory cells. FREEDMAN et al. (1992) cocultured FDC with B cells and observed an inhibitory action of FDC on B cell proliferation. In similar culture systems, BOSSELOIR et al. (1994) found the degree of B cell proliferation in the presence of FDC to vary according to the type of mitogenic stimulation. For example, B cells cultured alone but stimulated with *Staphylococcus aureus* proliferated better than cells grown in the presence of mitomycin C-treated FDC, whereas B cells stimulated with anti-Ig antibodies incorporated more tritiated thymidine when placed in contact with FDC; nonadherent lymphoid cells were used as controls (see Fig. 2). Thus, B cell multiplication may depend on the microenvironment created both by external influences and by phenotypically and thus locally different FDC types.

Germinal centers derive from polyclonal founder cells, i.e., activated B cells expressing unmutated antibodies in the initial dark zone (KUPPERS et al. 1993). Among these cells, only one to six precursors will develop into dominant clones populating the germinal centers (KROESE et al. 1987). The clones can reach populations as large as 2000 cells having gone through more than ten generations (KUPPERS et al. 1993).

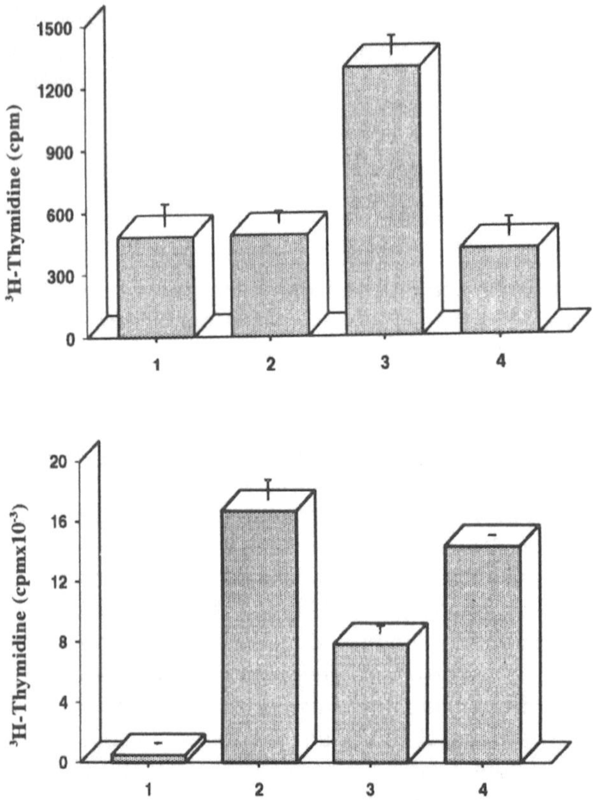

Fig. 2. Incorporation of tritiated thymidine into follicular dendritic cells (FDC) treated with mitomycin C (*1*), B cells cultured alone (*2*), mitomycin-treated FDC plus B cells (*3*), and B cells plus nonadherent mitomycin-treated cells (*4*). The cultures were maintained in RPMI–10% fetal calf serum containing either Staphylococcus aureus Cowan I (SAC; *bottom*) or anti-membrane immunoglobulin antibodies (anti-sIg; *top*)

7 Tolerance

Most published works are devoted to analyzing the stimulatory functions of the immune system and thus deal with activation, proliferation, and differentiation. Yet to maintain homeostasis, the immune defense system comprises suppression and tolerization mechanisms, i.e., means of downregulating certain responses.

The germinal centers are the theater of strong stimulatory processes. The curbing mechanisms have to be as strong to counteract this stimulation. Over, many potentially autoreactive B cell clones are produced there and must be tolerized. We must thus decipher these restraining and tolerizing systems in order to comprehend the physiology of the germinal centers.

After primary antigen injection, B cells go through a tolerization-sensitive window shortly after being stimulated by antigen, if they encounter that antigen is soluble form in the absence of concomitant T cell help (LINTON et al. 1991). Newly emerging memory B cells are highly susceptible to inactivation. For example, hapten-specific B cells can be blocked by recognition of a hapten on a carrier in the absence of T cells specific for oligopeptides of that carrier; the B cells consequently pass through a second tolerization-sensitive window (LINTON et al. 1991). The germinal center T cell appear to play an important role, since elimination of T cell help by means of a tolerogen (soluble antigen) significantly reduces memory B cell numbers (NOSSAL 1994).

The presentation of antibody–antigen complexes by FDC can lead to active suppression, since the cross-linking between Fc receptors and Ig produces the so-called Ig-dependent mouse B cell blockage (SINCLAIR 1969; PHILIPPS and PARKER 1985; MUTA et al. 1994). This mechanism may tolerize lymphoid cells expressing certain Fc receptors and thus act when certain Ig classes are secreted.

The idiotopic network (URBAIN 1986) may stimulate or inactivate. FDC can retain idiotype–anti-idiotype complexes (HEINEN, unpublished), so the germinal centers can be considered one of the sites where network regulation operates.

Relationships between FDC and the mantle zone cells are not clear. Mantle zone B cells are quiescent; some are tolerized and thus anergic (BEREK et al. 1991; KROEMER et al. 1993). For example, in transgenic mice carrying both a rearranged antilysozyme IgM/IgD transgene and a lysozyme transgene, tolerant lysozyme-reactive B cells persist within the follicular mantle zones but are absent from the splenic marginal zones. Selective accumulation of such B cells within the follicular mantle zones suggests unique physiological roles for this lymphoid micro-environment (MASON et al. 1992). An open question is whether anergic B cells can present self-antigens to T cells, thereby rendering them tolerent. FDC are present in primary follicles and mantle zones and may thus play a part in the tolerization events occurring there.

8 B Cell Survival

WYLLIE et al. (1984) have described apoptosis as an autonomous, cell suicide pathway that helps limit cell numbers. Morphologically distinct and highly "ritualistic" processes lead to cell disappearance. Cells dying by apoptosis display membrane blebbing, volume contraction, nuclear condensation, and activation of endonucleases that cleave DNA into nucleosome-length fragments. Programmed cell death occurs during various physiological events. A ligand on a given receptor, for example, can induce apoptosis, as do corticoids to thymocytes. The absence of ligands for receptors can also lead to apoptosis (the absence of nerve growth factor results in nerve degeneration). The mechanisms of programmed cell death also serve to accelerate cell disappearance when cells are injured or starved.

In the germinal centers, all of these mechanisms may operate. After Ig hypermutation, for example, cells with decreased antigen affinity enter apoptosis (absence of ligand fixation). In the dark zone, where cells divide at a high rate, cell death can result from replication errors or from missed DNA rearrangements during the isotype switch. Numerous tingible body macrophages are found in the dark zone, indicating massive cell degeneration in this area. Situations inside the germinal centers where ligand binding actively induces cell suicide are not known, but such events are suspected since programmed cell death can be triggered in lymphoid cells, for example by treatment with anti-Fas antibodies. Conversely, a lack of Fas or its decreased expression in transgenic mice leads to defective apoptosis and to autoimmune conditions resembling lupus erythematosus (MIYAWAKI et al. 1992; AKBAR et al. 1993). Fas antigen/APO-1 belongs to the family of cell surface proteins which includes the tumor necrosis factor (TNF) receptors, CD40, and the nerve growth factor receptor. Its expression is restricted on the one hand to activated T and B cells, but also to cells that harbor memory, such as CD45 RO$^+$ T cells and surface Ig (sIgD)-B cells (MIYAKAWA et al. 1992). Like TNF-α receptors I and II, Fas/Apo-1 induces apoptosis after binding of its ligand, in contrast to CD40 and the nerve growth factor (NGF) receptor, which ligand binding saves from cell death (TRAUTH et al. 1989). As with TNF, the biological responses triggered by the Fas/Apo-1 ligand may, in specific situations, promote activation, thus favoring life over death (ALDERSON et al. 1993).

Bcl-2 protects from apoptosis. Lymphoid cells express Bcl-2 abundantly, scarcely, or not at all (KORSMEYER 1992; AKBAR et al. 1993; LAGRESLE et al. 1993). This factor apparently functions as a homodimer. Interestingly, a homologous protein called Bax, also a homodimer, appears to stimulate entry into apoptosis (OLTVAI et al. 1993). Reduced Bcl-2 expression after immune stimulation may render cells prone to apoptosis. This could be a mechanism for controlling cell numbers by removing unwanted cells, for example after recovery from an infection or another stimulus (AKBAR et al. 1993). Overexpression of Bcl-2 in transgenic mice extends the secondary immune response and prolongs the survival of memory cells (NUNEZ et al. 1991). Such mice display numerous antibody-secreting cells 75 days after secondary immunization, by which time only baseline levels are detected in normal mice. Transgenic mice with Bcl-2 Ig minigenes develop diffuse large cell immunoblastic lymphoma after a latency period characterized by higher proliferation and increased small resting cell survival (STRASSER et al. 1990; McDONNELL and KORSMEYER 1991).

Inside the germinal centers, centroblasts and centrocytes located in the dark and lower light zones do not express Bcl-2 (LIU et al. 1991b; KORSMEYER 1992). These cells enter apoptosis when their affinity for antigen is lowered by Ig hypermutation (see above), but other inducer systems for B cell apoptosis act inside the germinal centers. The corresponding signals include stress hormones and decreased levels of growth factors or ligands due to competition between activated clones, suppressor cells, etc.

Since, 1984, our group has highlighted the positive effect of FDC on the survival of germinal center lymphoid cells. In organ cultures of dissected tonsil follicles, lymphoid cells show signs of degeneration very early. The mantle and germinal center cells die in large numbers, except those enveloped by FDC cytoplasmic extensions, which survive longer (CORMANN et al. 1986a). TSUNODA et al. (1990, 1992) and KOSCO and GRAY (1992) have also observed survival of lymphocytes in contact with isolated FDC. They stressed that only emperipolesis ensures survival. According to KOOPMAN and PALS (1992), LFA-1/ICAM-1 interactions then prevent programmed cell death.

After specific antigen stimulation outside the germinal centers, virgin B cells undergo a stepwise series of events enabling them to be activated, to proliferate, to hypermutate, to switch their isotype, and to express certain receptors. At each step, however, they are also subject to stringent environmental controls during which they risk elimination by apoptosis. FDC, by providing nonspecific antigenic signals, e.g., surface ligands, CD23, IL-1α or specific ones (via retained immune complexes), play a determining role during this evolutionary process (LIU et al.1989, 1991a, b).

Germinal center T cells, through the interaction of CD40 and CD40L, also play a role in rescuing B cells from apoptosis. We (BOUZAHZAH et al., to be published) have demonstrated the presence of mRNA for CD40L in CD4$^+$ CD57$^-$ cells from tonsillar follicles, and others (LEDERMAN et al. 1992; TSUBATA et al. 1993) report the ability of CD40L effectively to block B cell apoptosis in germinal centers.

The survival of memory B lymphocytes seems to require the continued presence of antigen arising from recurrent infections or from the antigen reservoir formed by the immune complexes retained of FDC (GRAY and SKARVALL 1988).

9 Immunoglobulin Isotype Switch

Specific immune responses are also called adaptive reactions, notably because Ig affinity and isotype switches increase during the primary responses. These switches produce the change of Ig classes and subclasses, thus optimizing the efficacy of their protective action. Roughly speaking, IgM act in the blood compartment, IgG in the blood and intercellular spaces, IgA in the extracorporal areas, and IgE at the surface of basophils, mast cells, and eosinophils. According to their capacity to activate the complement cascade or to bind to Fc receptors, these different Ig classes perform specialized tasks. In short, IgM are antimicrobial, IgG block toxins, IgA neutralize external agents, and IgE stimulate antiparasite defenses.

The isotype switch phenomenon is not fully understood. It occurs as early as lymphopoiesis, after IgM expression in virgin cells for formation of IgD, but it functions mainly during immunopoiesis, thus after contact with antigen, and depends on environmental parameters, such as T cells, microbial products, and nature of the antigen. Literature on this topic abounds, and space is lacking here for a complete overview. Studies have especially underlined the role of T cells, which control the Ig isotype switch via cytokines (Banchereau et al. 1992) or by direct contact via CD40L (Callard et al. 1993; Durandy et al. 1993; Lane et al. 1994), perhaps during B cell proliferation (Fridman 1993). For Hogkin et al. (1994), the signal is given by cytokines before the start of DNA synthesis, but the isotype switch itself occurs during the S phase.

Follicular B cells express the various Ig classes, but the latter vary according to the tissular location: in the lymph nodes and tonsils, germinal centers are mainly composed of IgG^+ B cells; in mucosa-associated lymphoid tissue (MALT), they contain IgA^+ cells (Butcher et al. 1982; Matthews and Basu 1982. Brandtzaeg 1989). It is unclear which specific environments promote isotype switching. Butcher et al. (1982) suggest it happens inside the follicles, but other places have been proposed: the T-dependent zones during virgin B cell recruitment, the mucosa after settlement of B memory cells and during terminal differentiation to Ig-secreting cells.

Participation of FDC in the Ig class switch has not been demonstrated, but it is likely since FDC retain immune complexes composed of various Ig classes. Immune complexes can act via Fc receptors on B, T, or other cells and are thus liable to induce opposite effects: Fc receptors on B cells generally transmit negative, blocking intracellular signals (Sinclair 1969; Abbas and Unanue 1975; Philipps and Parker 1985; Rigley et al. 1989; Muta et al. 1994), while on T cells, Fc receptors are expressed only during activation (Sandor and Lynch 1993) and stimulate or suppress according to the bound Ig isotype. On mast cells, Fc receptors induce degranulation and, among others, IL-4 secretion, which induces formation of new IgE. On dendritic cells, Fc receptors promote antigen capture and presentation and hence T cell and then B cell stimulation followed by Ig secretion. On macrophages, Fc receptrors induce production of prostaglandins which inhibit T cells, etc.

The nature of the immune complexes retained by FDC probably reflects the Ig composition in the circulating blood. In the blood, concentrations of the different Ig classes are maintained within narrow limits. The mechanisms controlling Ig secretion are unknown, but the complexes on FDC might play a part therein. Wang et al. (1994) demonstrated that T-independent antigens of type II are retained on FDC and induce a typical germinal center reaction, though without the classical Ig isotype switch. Thus immune complex retention does not automatically induce switching.

10 Affinity Selection

Siskind and Benacerraf (1969) postulated, in accordance with Burnet's clonal selection hypothesis, that antibody affinity increases in time because precursor cells have to compete for decreasing amounts of antigen, so that ultimately only cells with high-affinity receptors can capture enough antigen to be triggered.

This view has turned out to be partly correct. Connections between memory cell formation, germinal centers, and affinity selection were established notably after the observation that FDC retain antigens for long periods, even when antigen has disappeared from other areas (NOSSAL et al. 1965; TEW et al. 1979). We are beginning to understand the mechanisms underlying affinity selection. Among these, somatic hypermutation and decreased membrane Ig expression by germinal center B cells appear essential.

Ig hypermutation and selection take place in the germinal centers (JACOB et al. 1993; BEREK and ZIEGNER 1993). Mutations are introduced into rearranged regions of transcriptionally active loci on the heavy and light Ig chains (SABLITZKY et al. 1985), at the rate of 10^{-3} mutations per base pair, per cell, and per generation (CLARKE et al. 1985). These mutations preferentially accumulate around the first complementary determinant V region and are mainly single nucleotide substitutions. Transitions are preferred to transversions, although deletions and insertions have been revealed (ALLEN et al. 1987; BEREK and MILSTEIN 1987; MANSER et al. 1987; LEBECQUE and GEARHART 1990; BETZ et al. 1993). These mutations occur asymmetrically within a 2-kb pair region whose 5′ boundary is defined by the Vh promoter sequence (LEBECQUE and GEARHART 1990). The mechanism of Ig hypermutation is unclear: mutations appear at random, but strand biases and mutational hot spots have been found; they differ from mutations arising during meiosis (GOLDING et al. 1987). The relative mutation frequency is highest in the complementarity-determining regions (CDR) that encode the paratope (CUMANO and RAJEWSKY 1986). Recurrent key mutations in H and L chain V regions increase the affinity for the paratope (ALLEN et al. 1987; WEISS et al. 1992). Mutations that abolish or diminish the capacity of antibody to bind antigen are thought to result in apoptosis (LIU et al. 1989). JACOB et al. (1993) have analyzed the kinetics of V region mutation and selection in germinal center B cells: although germinal centers appeared by day 4 after immunization, mutations were not observed until day 8; thereafter, point mutations favoring asymmetrical transversions accumulated until day 14. During this period, the mutant B lymphocytes were subject to strong phenotypic selection, as inferred from the progressively biased distributions of mutations within the Ig variable region, the disappearance of crippling mutations, a decreased relative clonal diversity, and an increasingly restricted use of canonical gene segments. The same authors also conclude that the period of most intense selection on germinal center B cell populations precedes significant levels of mutation. It may represent an important physiological mechanism restricting the entry of the B cells into the memory pathway. According to this view, selection operates even before the mutation process is initiated, early after antigen administration. This underlines the fact that, besides Ig hypermutations, other bases for selection exist. Germinal center B cells, especially centroblasts in the dark zone, are reputed to express low levels of mIg. They are much more prone to affinity selection than are other B cells expressing high levels of mIg. This view is supported by the recent paper by GEORGE et al. (1993), who hypothesize that downregulation of mIg following B cell activation has evolved to assist selection of B cell clones. ROES and RAJEWSKI

(1993), studying the role of IgD in transgenic IgD$^-$ mice, found that IgD are lost after activation. They suggest that this may be a part of the affinity selection mechanism. IgD do seem to improve the avidity of B cells for antigens, so their disappearance decreases the chances of survival of B cells with low-affinity mIg. George et al. (1993) also conclude that T-independent antigens, usually large in size and multimeric, easily stimulate B cells by cross-linking mIg, even when the density of mIg is low. These antigens do not favor a form of affinity selection based on low membrane density.

Other processes may be coupled to affinity selection. We hypothesize that B cells which have recognized antigen and escaped apoptosis may exert a dominance on other cell lines, rendering them unable to become activated in the same area. This cell dominance would also favour the oligo- or monoclonality of the germinal centers, a fact observed by several authors (Kroese 1987; Liu et al. 1991a, b). According to Jacob et al. (1993), the period of most intense selection on germinal center cell populations occurs between days 6 and 8 of the response; this coincides with the IgM–IgG switch in germinal centers and with the initiation of somatic hypermutation. Thus, perhaps someone will discover a relationship between affinity selection and the isotype switch, although Manser (1989, 1990), has found no link between switch and mutation.

Little is known about factors initiating and controlling Ig hypermutation. Mutations appear mainly within centroblasts, probably during the mitotic cycles, at the rate of one (Berek and Ziegner 1993) or more (Kuppers et al. 1993) mutations per two cell cycles. Manser (1990), however, believes that mutations and mitoses proceed independently. Proliferation appears necessary but not sufficient. Both T cells and a germinal center microenvironment are further required to initiate and regulate the hypermutation process. Apparently, periods of rapid mutation alternate with periods of mutation-free growth (Kepler and Perelson 1993).

Ig hypermutation starts after a 6-day latency period (Jacob et al. 1993). There may be a link between immune complex retention of FDC, mutation, and affinity selection. No direct action of FDC on Ig hypermutation in B cells has been demonstrated, but such action is imaginable, since mutations appear to be restricted to the germinal centers (Berek et al. 1991; Kepler and Perelson 1993) and the mutation process starts at about the same time as immune complex retention by FDC. The proposition that B cell activation in the T-dependent zone gives the starting signal for hypermutation is hardly tenable, since the mutations occur mainly after day 6. The link between immune complex retention and Ig hypermutation is strengthened by experiments performed by Kunkl and Klaus (1981), in which hapten-specific B memory cells were induced by priming mice with soluble or alum-precipitated dinitrophenol-keyhole limpet hemocyanin (DNP-KLH) plus *Bordetella pertussis* or DNP-KLH–anti-DNP antibody complexes at equivalence. Cells from complex-treated mice gave a substantial adoptative IgG response 5 days after priming, whereas cells from mice treated with antigen and adjuvant gave no comparable response until day 14. The relative affinity of the adoptative secondary IgG response induced by priming with complexes was already maximal by day 6. In contrast, the response of memory cells from mice given antigen in alum increased in affinity between day 6 and 23 after priming. These authors suggested that trapping of antigen–antidoby complexes induces the formation of germinal centers and that selective triggering of high-affinity precursor cells gives rise to functional B memory cells.

FDC might participate in affinity selection at different levels. By creating an adequate microenvironment, they may induce homing of lymphoid cells engaged in hypermutation and favor initiation of this process. The role of the antigen–antibody complexes during clonal selection is widely accepted. One may object that these complexes are not all related to the stimulating antigen, since they are retained in a nonspecific manner (KLAUS et al. 1980; RADOUX et al. 1984). However, this antigen diversity may be advantageous because it provides germinal center B cells with a very large spectrum of antigens.

11 B Memory

The best hallmark of a memory B cell may well be the somatically mutated immunoglobulin genes (MACKAY 1993). Other markers for memory are clonal enlargement, the Ig isotype switch, loss of Fc receptor, and changes in selectins (HEINEN et al. 1990). A connection between memory cell formation and the germinal center reaction was suggested early on by THORBEKE et al. (1962). The determining role that the immune complexes retained by FDC play in memory cell formation was shown long ago by PAPAMICHAIL et al. (1975) and KLAUS et al. (1977). These authors found a correlation between the absence of immune complex trapping after complement depletion by cobra venom and the absence of memory induction. Furthermore, periodical contact with antigen retained by FDC seems necessary for memory retention. When cells are transferred from an immune animal to a naive one in the absence of antigen, the memory cell population is lost within 12 weeks (GRAY 1988), but when antigen is provided and attaches to the FDC, the memory response is sustained much longer (GEORGE and CLAFIN 1992).

The immune complexes retained on FDC can contain an excess of antigen or antibody. At the time the humoral immune response is initiated, one can suspect antigen to be in excess, whence the possibility of B cell activation. Later on, when antibodies are formed, free antibody will compete with the membrane-bound Ig on B cells and thereby diminish the intensity of the immune response. TEW et al. (1980) proposed that the cyclical variation of the antigen to antibody ratio regulates the germinal center reaction. One might also speculate the antibody in excess masks the antigen, thus curbing or stopping affinity selection, a part of the germinal center reaction.

Several complement factors form bridges between FDC and B cells, thereby giving inhibitory or stimulatory signals (ERDEI et al. 1985; HEYMAN 1990; VAN ROOIJEN 1991; VAN NOESEL et al. 1993). These contradictory data are rendered even more complex by the fact that immune complexes are also retained by Fc receptors on mouse (RADOUX et al. 1985b) and human (HEINEN et al. 1984) FDC. Some authors contest the presence of Fc receptors on human FDC. YAMAKAWA and IMAI (1992) and YOSHIDA et al. (1993) found FDC to differ functionally according to differences in complement and/or Fc receptor expression. Thus nowadays no clear interpretation can be given as to the relative functions of antigen, complement, Fc regions, and the antigen to antibody ratio inside the germinal centers. Experiments based on the use of cobra venom can be misleading, since this product induces heavy complement activation followed by an overload of complement fragments capable of masking Fc and other receptors (ERDEI et al. 1992).

The nature of the germinal center precursor cells is hardly discussed. Virgin IgM^+ IgD^+ cells can populate germinal centers (ROZING et al. 1978; ENRIQUEZ-RINCON et al. 1984; BAZIN et al. 1985). B cells fractionated into IgD^+ and IgD^- populations gave rise, in each case, to germinal center precursors (SEIJEN et al. 1988; VON DER HEIDE and HUNT 1990).

LINTON et al. (1991) suggest that these precursors may reside in the $J11D^-$ poor population. J11D (CD24) appears as a cell adhesion molecule expressed on immature B and T cells (KADMON et al. 1992). According to TSIAGBE et al. 1992), the J11D-rich cells may terminally differentiate into antibody-forming cells and thus never reach the centers. We have found (MANCINI et al., to be published) germinal center cells expressing high J11D, especially the dark zone. Since these cells are very sensitive to apoptosis and do not express homing receptors, they may degenerate during the experimental procedure or be unable to home to centers or to migrate because they adhere to other cells. Thus, we still need cell markers to clearly identify the germinal center precursors.

12 Germinal Center T Cells

Germinal centers develop in response to T-dependent antigens. Nu/nu or thymectomized mice exhibit practically no germinal centers (JACOBSON et al. 1974; GROSCURTH 1980b; KROESE 1987), but FDC exist in their primary follicles in a poorly differentiated state (GROSCURTH 1980b). Since T cells are present in germinal centers (GUTMAN and WEISSMAN 1972) and establish contacts with FDC (HEINEN et al. 1989; LOUIS et al. 1990), one might suspect reciprocal influences between the two cell types. COSGROVE et al. (1991) discovered a $CD4^+$ T cell type in the lymph follicles of mice lacking MHC class II molecules, whereas the other peripheral lymphoid tissues were devoid of $CD4^+$ cells. These quite specific cells are relatively large in size, T cell receptor $\alpha\beta^+$, and express high levels of CD44.

Germinal center T cells represent 5%–20% of the lymphoid cell population; most are $CD4^+$ cells and $CD8^+$ cells are rare (RITCHIE et al. 1983; SI and WHITESIDE 1983; KROESE et al. 1985; POPPEMA et al. 1989; TSUNODA et al. 1990). Among the $CD4^+$ cells, a majority express the CD57 antigen (MORI et al. 1983; SI and

WHITESIDE 1983; OKADA et al. 1988). These CD4⁺ CD57⁺ cells constitute a peculiar cell type: they express T cell receptor-α and -β, CD45RO, and CD69 but not Leu8, CD45/RA, CD11b, CD25, CD71, or HLA-DR (PORWIT-KSIAZEK et al. 1983; PIZZOLO et al. 1984; SI and WHITESIDE 1983; VELARDI et al. 1986b). They are mainly located in the germinal centers (MORI et al. 1983; RITCHIE et al. 1983; SI and WHITESIDE 1983), especially the light zone (BOUZAHZAH et al. 1993). Ultrastructural studies show that they are medium-sized cells with no or only small granules (BOUZAHZAH et al. 1993) and that they differ from typical NK cells even though they express the CD57 antigen. These CD4⁺ CD57⁺ cells are not fully activated cells, being CD25⁻, CD71⁻, and HLA-DR⁻. They appear, rather, to be preactivated (CD69⁺).

In vitro tests have shown that CD4⁺ CD57⁺ cells do not perform the usual functions of classical T or NK cells: they are not cytotoxic (SI and WHITESIDE 1983; VELARDI et al. 1986a; BOUZAHZAH et al., to be published); they do not produce B cell growth factors (VELARDI et al. 1986a), IL-2, IL-4, interferon (IFN)-α,or TNF-α (BOWEN et al. 1991), or IL6 (BOSSELOIR et al. 1989); they do not support B cell proliferation or differentiation (VELARDI et al. 1986a; Fig. 3; BOUZAHZAH et al., to be published). There are contradictory reports as to whether they produce IL-4: BUTCH et al. (1993) found mRNA for IL-4, whereas BOSSELOIR et al. (1991) did not detect this cytokine in germinal centers by in situ hybridization, data confirmed by histochemistry by HOEFAKKER et al. (1993).

CD4⁺ CD57⁺ cells do not divide immediately after isolation, and even the presence of FDC cells does not induce their proliferation (Fig. 4; BOUZAHZAH et al.

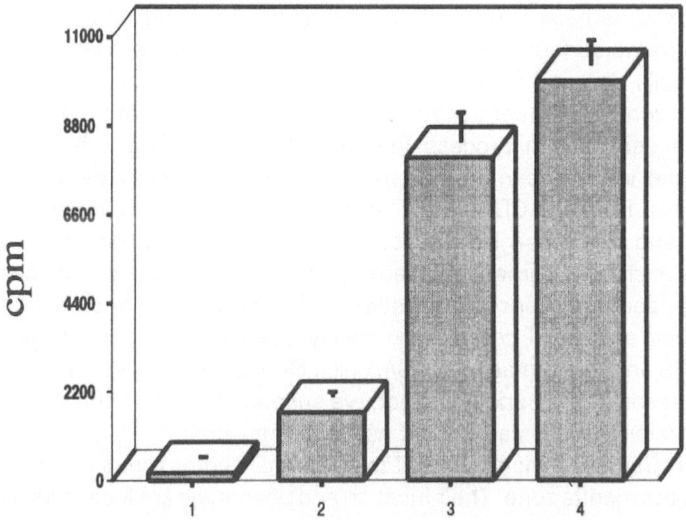

Fig. 3. Proliferation capacity of tonsillar B cells cultured in contact with CD4⁺ T cells. Tritiated thymidine incorporation was measured after 2 days of culture in the presence of phytohemagglutinin (PHA). All T cells were treated with mitomycin C before coculturing to prevent them from multiplying. *1*, B cells alone; *2*, CD57⁺ T cells; *3*, B ⁺CD57⁻ T cells; *4*, B and unsorted CD4⁺ cells. The T to B cell ratio was 3:1. CD4⁺ CD57⁺ cells induced B cell proliferation but weakly in comparison to CD4⁺ CD57⁻ cells. *cpm*, counts per minute

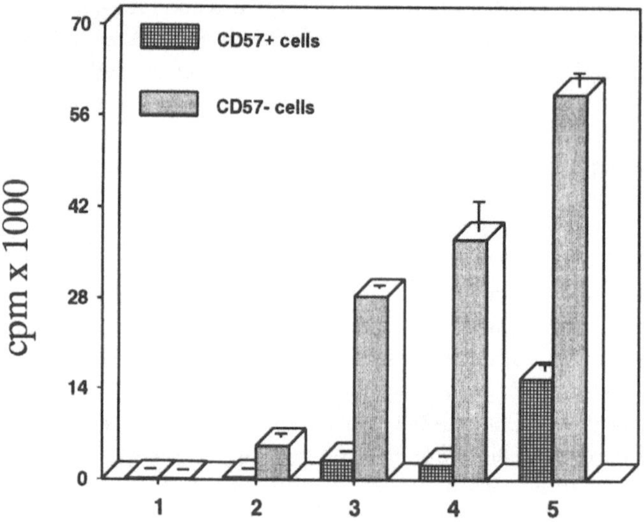

Fig. 4. Proliferation test of CD4⁺ CD57⁺ and CD4⁺ CD57⁻ T cells. Purified tonsillar CD4⁺ T cells were separated into CD57⁻ and CD57⁺ populations and maintained in culture for 2 days before testing their capacity to incorporate tritiated thymidine. *1*, Culture medium without mitogens; *2*, phytohemagglutinin (PHA); *3*, PHA and interleukin IL-2; *4*, concanavalin A (ConA); *5*, ConA and IL-2. Under all experimental conditions, CD57⁻T cells proliferated better than CD57⁺ T cells. *cpm*, counts per minute

1994). In migration tests, T cells, especially CD4⁺ CD57⁺ T cells, are attracted by FDC or their culture supernatants (BOUZAHZAH et al., to be published). In adhesion tests, they attach to the surfaces of FDC clusters as readily as B cells do.

Since these cells do not function like classical T cells and since they are located in the light zone where centrocytes are present, we speculate that they play a role in the deactivation process through which centroblasts become centrocytes and the latter become small, recirculating, B memory cells. The relationship between FDC and CD4⁺ CD57⁺ cells in such a process in unclear. Since FDC attract and bind these T cells (BOUZAHZAH et al., to be published), they perhaps create a microenvironment favorable to them, to expression of their typical phenotype, and to the encounter between T and B cells. In rats, KROESE et al. (1985) detected a similar T cell subtype mainly located in the light zone and expressing the ER3 antigen, a marker of Ts/c cells. During the humoral immune responses, suppressor T cells appear and act; we suspect that the CD57⁺ T cell population might belong to this category of negative regulators.

Classical CD4⁺ CD57⁻ T cells occupy all the germinal center zones and also, to a lesser extent, the mantle zone. Their function and phenotype are less studied. Apparently, they are also nonactivated, nondividing T cells of medium size (LOUIS et al. 1989). On the basis of many experiments, KOSCO et al. (1988, 1992; KOSCO and GRAY 1992) have described T cells which exert a positive action on B cells cultured in the presence of FDC. Perhaps the cells they observed are the germinal center CD57⁻ T cells. According to GRAY et al. (1991), FDC do not express MHC class II molecules, but they can take them up from B cells and present antigens to T cells. We do not hold this view, since germinal center T cells appear only in a preactivated stage and do not divide. We believe antigen retained on FDC can be picked up by B cells, processed, and presented to T cells located in the T-dependent area.

Recent experiments carried out on transgenic mice overexpressing mouse CTLA4–human-α_1 protein shed light on the role of the germinal center T cells. In such animals, the germinal centers do not form and this correlates with impaired switching to the IgG isotype, reduced somatic mutation and selection, and reduced antibody production (LANE et al. 1994). These experiments make sense in the light of work done by MUNRO et al. (1994) showing that lymph follicles contain cells expressing either B7 or CD28. Thus, contacts via surface molecules appear to control the cellular phenomena inside the germinal centers, where cytokines do not appear as key elements (HEINEN et al. 1991b, c).

13 Conclusions

Many facts are known about FDC and their relationships with germinal center lymphoid cells. However, we cannot always discern causes form consequences. For instance, do FDC decide the course of B memory cell maturation or do lymphoid cells induce FDC to differentiate, thus laying the scene for a humoral immune response?

To clarify the situation, we have tried to outline the chronology of a T-dependent immune response after primary antigen stimulation. This brief chronological description should be read with caution, as the published data on which it is based come from experiments which used different antigens and adjuvants and which differed in the sex, age, and species of the animals used (LENNERT 1978; KROESE et al. 1987, 1991; KOSCO et al. 1989; BEREK et al. 1991; JACOB et al. 1993). Figure 5 tentatively summarizes the phenomena occurring during a germinal center reaction. For simplicity's sake, the case of a draining lymph node was chosen, but the events in lymph node germinal centers are similar to those of the spleen or MALT system:

1. *Antigen recognition by B cells.* Circulating B cells enter the T-dependent zone via high endothelial venules and encounter native antigen imported by afferent lymph. Antigen-laden B cells apparently also enter via the afferent lymphatics.
2. *T cell activation.* Antigen-presenting dendritic cells immigrate via the lymphatics and settle in the T-dependent zone, yielding interdigitating cells which activate T cells.
3. *B cell activation.* B cells, stimulated by antigen and activated T cells, reach the lymph follicles, where they enlarge and appear as centroblasts gathered in small nodules, the first signs of germinal center formation.
4. *B cell proliferation.* After a latency period, the centroblasts divide. This multiplication continues through the entire period of the germinal center reaction.

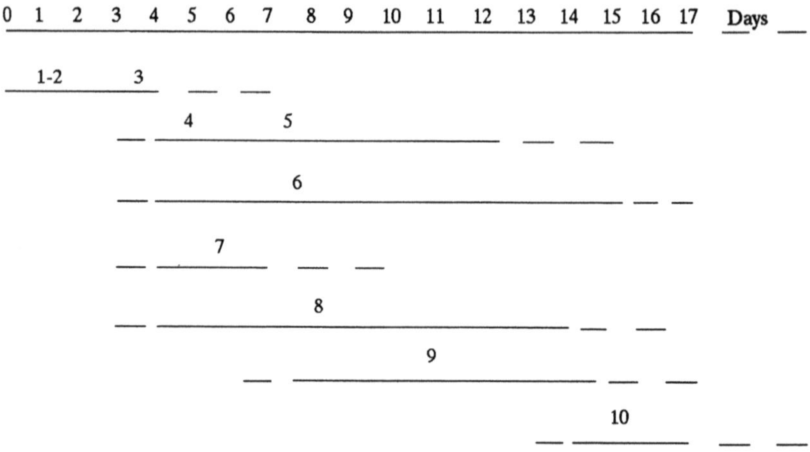

Fig. 5. Chronological events of a primary germinal center reaction in a lymph node. For explanation see text

5. *Compartmentalization of the germinal centers.* Several days after antigen injection, centrocytes appear at one pole, the light zone of the germinal centers, and drive back the centroblasts, which then form the dark zone. At the same time immune complexes are retained in a crescent in the upper light zone, along FDC located in the vicinity of the centrocytes. Further subdivisions have been described (HARDIE et al. 1993).
6. *Apoptosis.* Concomitantly with the appearance of mitoses, pycnotic cells are observed and tingible body macrophages develop, especially in the dark zone. The reasons for apoptosis are multiple: missed rearrangements during the isotype switch, crippling somatic mutations, no repair during mitosis, etc.
7. *Proplasmoblast emigration.* During the first few days of germinal center formation, proplasmoblasts seem to emigrate from the follicles and to enter the medullary zone. During this journey along blood vessels, these cells differentiate into typical plasma cells which accumulate in the medullary cords. This apparently only lasts a few days (SIMAR and WEIBEL 1979; KOSCO et al. 1989).
8. *Isotype switch.* The induction signal for the Ig isotype switch is given, according to some authors, during B cell stimulation in the T-dependent area. It appears to occur during multiplication inside the germinal centers. However, the switch might also happen outside the follicles, for example in the mucosa.
9. *Affinity maturation.* Somatic mutations apparently occur in dividing centroblasts by day 8 and accumulate until day 14 (JACOB et al. 1993); antigen selection concerns both centroblasts and centrocytes, and thus both populations exhibit survival or apoptosis.
10. *End of the germinal center reaction.* Centroblasts stop dividing, the number of lymphoid cells decreases, and the centroblasts disappear before the centrocytes.

In this review we discuss the role of FDC at different levels: they can emit chemoattractants for B and T cells, allow selective cell attachment, and improve survival, proliferation, and positive selection. They apparently reduce the capacity of B cells to transform locally to plasma cells (CORMANN et al. 1986b), probably via the immune complexes they retain (KÖLSCH et al. 1983). This retention is not required to initiate the development of germinal centers (KROESE et al. 1986, 1991). As related to the Ig isotype switch and somatic mutation, the role of FDC is hypothetical. Other functions are suspected in normal or pathological conditions, e.g., B cell deactivation or retention of nonspecific activators (lipopolysaccharides, superantigens, viruses); much work is needed to elucidate all the functions of follicular dendritic cells.

Acknowledgments. We are grateful to Dr. T Defrance (Lyon, France) and Dr. R. Tsunoda (Fukushima, Japan) for fruitful discussions on the topics presented here.

References

Abbas AK, Unanue ER (1975) Interrelationship of surface immunoglobulin and Fc receptors on mouse B lymphocytes. J Immunol 115: 1665–1671

Akbar AN, Salmon M, Savill J, Janossy G (1993) A possible role for bcl-2 in regulating T-cell memory—a balancing act between cell death and survival. Immunol Today 14: 526–532

Alderson MR, Armitage RJ, Maraskovsky E et al (1993) Fas induces activation signals in normal human T lymphocytes. J Exp Med 178: 2231–2235

Allen D, Cumano A, Dildrop R et al (1987) Timing, genetic requirements and functional consequences of somatic hypermutation during B-cell development. Immunol Rev 5:

Ashman R (1990) B lymphocyte activation: the transferrin receptor as a prototype intermediate activation molecule. J Lab Clin Med 116: 759–765

Ashman RF (1984) Lymphocyte activation. In: Paul WE (ed) Fundamental immunology. Raven, New York, pp 267–302

Bachman MF, Hoffman-Rohrer U, Kündig TM, Bürki K, Hengartner H, Zinkernagel RM (1993) The influence of antigen organization of B cell responsiveness. Science 262: 1448–1451

Bancherau J, Rousset F (1992) Human B lymphocytes: phenotype, proliferation and differentiation. Adv Immunol 52: 125–262

Bazin H, Platteau B, MacLennan IC, Johnson GD (1985) B-cell production and differentiation in adult rats. Immunology 54: 79–88

Berek C, Milstein C (1987) Mutation drift and repertoire shift in the maturation of the immune response. Immunol Rev 96: 23–47

Berek C, Ziegner M (1993) The maturation of the immune response. Immunol Today 14: 400–404

Berek C, Berger A, Apel M (1991) Maturation of the immune response in germinal centers. Cell 67: 1121–1129

Betz GA, Rada C, Pannell R, Milstein C, Neuberger MS (1993) Passenger transgenes reveal intrinsic specificity of the antibody hypermutation mechanism: clustering polarity specific hotspots. Proc Natl Acad Sci USA 90: 2385–2389

Bosseloir A, Dehooghe-Peeters E, Heinen E et al. (1989) Production of Interleukin-6 in human tonsils by in situ hybridization. Europ J Immunol 19: 2379–2382

Bosseloir A, Hooghe-Peters E, Heinen E et al. (1991) IL6 and IL4: localization and production in human tonsils. In: Imhof BA, Berrih-Aknin S, Ezine S (eds) Lymphatic tissues and in vivo immune responses. Dekker, NewYork, pp 315–319

Bosseloir A, Heinen E, Defrance T, Bouzahzah F, Antoine N, Simar LJ (1994a) Immunol Lett 42: 49–54

Bosseloir A, Antoine N, Heinen E, Defrance T, Schuurman H, Simar LJ (1994b) Expression and function of DRC1 antigen. Adv Exp Med Biol (in press)

Bouzahzah F, Heinen E, Antoine N, Simar LJ (1993) Ultrastructure of $CD57^+$ cells isolated from human tonsils and blood. Eur J Morphol 31: 82–86

Bouzahzah F, Basseloir A, Heinen E, Simar LJ (1994) The germinal center CD4+ CD57+ T cells act differently on B cells than classical T helper cells. Dev Immunol (in press)
Bowen MB, Butch AW, Parvin CA, Levine A, Nahm MH (1991) Germinal center T cells áre distinct helper-inducer T cells. Hum Immunol 31: 67–75
Brandtzaeg P (1989) Overview of the mucosal immune system. In: Mestecky J, McGhee JR (eds) New strategies for oral immunization. Springer, Berlin Heidelberg New York, pp 13–25 (Current topics in microbiology and immunology, vol 146)
Brandtzaeg P, Nilssen DE, Rognum TO, Thrane PS (1991) Ontogeny of the mucosal immune system and IgA deficiency. Gastroenterol Clin North Am 20: 397–439
Braun M, Heinen E, Cormann N, Kinet-Denoël C, Simar LJ (1987) Influence of immunoglobulin isotypes and lymphoid cell phenotype on the transfer of immune complexes to FDC. Cell Immunol 107: 99–106
Brown JC, Harris G, Papamichail M, Slivjie VS, Holborow EJ (1973) The localization of aggregated human gammaglobulin in the spleen of normal mice. Immunology 24: 955–968
Butch AW, Chung GH, Hoffman JW, Nahm MH (1993) Cytokine expression by germinal center cells. J Immunol 150: 39–47
Butcher EC, Rouse RV, Coffman RL, Nottenbourg CN, Hardy RR, Weissman IC (1982) Surface phenotype of Peyer's patch germinal center cells: implications for the role of germinal centers in B cell differentiation. J Immunol 129: 2698–2707
Callard RE, Armitage RJ, Fanslow WC, Spriggs MK (1993) CD40 ligand and its role in X-linked hyper-IgM syndrome. Immunol Today 14: 559–564
Cebra J, Schrader C, Shroff K et al. (1991) Germinal centres and the immune response—discussion. Res Immunol 142: 275–282
Celada F (1971) The cellular basis of immunologic memory. Prog Allergy 15: 223–267
Chen L, Linsley PS, Hellström KE (1993) Costimulation of T cells for tumor immunity. Immunol Today 14: 483–486
Clark EA, Lane PJL (1991) Regulation of human B-cell activation and adhesion. Annu Rev Immunol 9: 97–127
Clark EA, Grabstein KH, Shu GL (1992) Cultured human follicular dendritic cells—growth characteristics and interactions with lymphocytes-B. Immunol 148: 3327–3335
Clarke SH, Huppi K, Ruezinsky D, Staudt L, Gerhard W, Weigert M (1985) Inter- and intra-clonal diversity in the antibody response to influenza hemagglutinin. J Exp Med 161: 687–692
Cormann N, Heinen E, Kinet-Denoël C, Braun M, Simar LJ (1986a) Influence de la cellule folliculaire dendritique sur la survie des lymphocytes in vitro. CR Soc Biol 180: 218–223
Cormann N, Lesage F, Heinen E, Schaaf-Lafontaine N, Kinet-Denoël C, Simar LJ (1986b) Isolation of follicular dendritic cells from human tonsils or adenoids. V. Effect on lymphocyte proliferation and differentiation. Immunol Lett 14: 29–35
Cormann N, Heinen E, Kinen-Denoël C. Simar LJ (1989) Isolation of follicular dendritic cells from human tonsils or adenoids. IV. In vitro culture. Adv Exp Med Biol 237: 171–176
Cosgrove D, Gray D, Dierich A et al. (1991) Mice lacking MHC c1II molecules. Cell 66: 1051–1066
Cumano A, Rajewsky K (1986) Clonal recruitment and somatic mutation in the generation of immunologic memory to the hapten NP. EMBO J 5: 2459–2463
Dijkstra CD, van Tilburg NJ, Döpp EH (1982) Ontogenetic aspects of immune-complex trapping in the spleen and the popliteal nodes of the rat. Cell Tissue Res 223: 545-552
Dijkstra CD, Kamperdijk EWA, Döpp EA (1984) The ontogenic development of the follicular dendritic cells. An ultrastructural study by means of intravenously injected horseradish peroxidase (HRP)-anti HRP complexes as marker. Cell Tissue Res 236: 203–206
Durandy A, Schiff C, Bonnefoy JY et al. (1993) Induction by anti-CD40 antibody or soluble CD40 ligand and cytokines of IgG, IgA and IgE production by B-cells from patients with X-linked hyper-IgM syndrome. Eur J Immunol 23: 2294–2299
Enriquez-Rincon F, Andrew E, Parkhouse RME, Klaus GGB (1984) Suppression of follicular trapping of antigen-antibody complexes in mice treated with anti-IgM or anti-IgG antibodies from birth. Immunology 53: 713–720
Erdei A, Füst G, Gergely J (1991) The role of C3 in the immune response. Immunol Today 12: 332–337
Everett NB, Tyler RW (1967) Radioautographic studies of reticular and lymphoid cells in germinal centers of lymph nodes. In: Cottier H, Odartchenko N, Schindler R, Congdon C (eds) Germinal centers in immune response. Springer, Berlin Heidelberg New York, pp 145–151
Freedman AS, Nadler LM (1991) Cellular interactions within the germinal centre. Res Immunol 142: 232–236

Freedman AS, Munro JM, Rhynhart K, Schow P, Daley J, Lee N, Svahn J, Elisco L, Nadler LM (1992) Follicular dendritic cells inhibit human lymphocyte-B proliferation. Blood 80: 1284–1288

Fridman WH (1993) Regulation isotypique. In: Bach JF (ed) Traité d'immunologie. Flammarion, Paris, pp 538–548

George J, Clafin L (1992) Selection of B cell clonal and memory B cells. Semin Immunol 4: 11–17

George J, Penner SJ, Weber J, Berry J, Claflin JL (1993) Influence of membrane Ig receptor density and affinity on B-cell signaling by antigen—implications for affinity maturation. J Immunol 151: 5955–5965

Gerdes J, Stein H, Mason DY, Ziegler A (1983) Human dendritic reticulum cells of lymphoid follicles: their antigenic profile and their identification as multinucleated giant cells. Virchows Arch [B] 42: 161–172

Golding BG, Gearhart PJ, Glickman BW (1987) Patterns of somatic mutations in immunoglobulin variable genes. Genetics 115: 169–176

Gray D (1988) Recruitment of virgin B cells into an immune response is restricted to activation outside lymph follicles. Immunology 65: 73–79

Gray D (1989) Memory B cells but not virgin B cells are activated in germinal centers. Adv Exp Med Biol 237: 209–214

Gray D, Skarvall H (1988) B-cell memory is short-lived in the absence of antigen. Nature 336: 68–69

Gray D, MacLennan ICM, Lane PJL (1986) Virgin B cell recruitment and the lifes span of memory clones during antibody responses to 2,4-dinitrophenyl-hemocyanin. Eur J Immunol 16: 641–648

Gray D, Kosco M, Stockinger B (1991) Novel pathways of antigen presentation for the maintainance of memory. Int Immunol 3: 141–148

Groscurth P (1980a) Non-lymphoid cells in the lymph node cortex. I. Morphology and distribution of the interdigitating cells and the dendritic reticular cells in the mesenteric lymph node of the adult ICR mouse. Pathol Res Pract 169: 212–234

Groscurth P (1980b) Non-lymphatic cells in the lymph node of the mouse. Pathol Res Pract 169: 255–268

Gutman G, Weissman I (1972) Lymphoid tissue architecture: experimental analysis of the origin and distribution of T and B cells. Immunology 23: 465–471

Hanna MG, Szakal AK (1968) Localization of 125I-labelled antigen in germinal centers of mouse spleen: histologic and ultrastructural autoradiographic studies of the secondary immune responses. J Immunol 101: 949–962

Hardie D, Johnson G, Khan M, MacLennan I (1993) Quantitative analysis of molecules which distinguish functional compartments within germinal centers. Eur J Immunol 23: 997–1004

Heinen E, Bosseloir A (1994) Follicular dendritic cells: whose children? Immunol Today 15: 201–205

Heinen E, Radoux D, Kinet-Denoël C, Simar LJ (1983) Colloidal gold, an useful marker for antigen localization on follicular dendritic cells. J Immunol Methods 59: 361–368

Heinen E, Lilet-Leclercq Ch, Mason DY et al. (1984) Isolation of follicular dendritic cells from human tonsils and adenoids. II. Immunocytochemical characterization. Eur J Immunol 14: 267–273

Heinen E, Radoux D, Kinet-Denoël C, De Mey J, Moeremans M, Simar LJ (1985) Isolation of follicular dendritic cells from human tonsils and adenoids. III. Analysis of their Fc receptors. Immunology 54: 777–784

Heinen E, Braun M, Coulie PG et al. (1986) Transfer of immune complexes from lymphocytes to follicular dendritic cells. Eur J Immunol 16: 167–172

Heinen E, Braun M, Louis E et al. (1989) Interactions between follicular dendritic cells and lymphoid cells. Adv Exp Med Biol 287: 181–184

Heinen E, Bosseloir A, Cormann N, Kinet-Denoël C (1990) Microenvironments during antigen stimulation. In: Sorg C (ed) Molecular biology of B cells developments. Karger, Basel, pp 24–60

Heinen E, Cormann N, Tsunoda R, Kinet-Denoël C, Simar LJ (1991a) Ultrastructural and functional aspects of FDC in vitro. In: Racz P, Dijkstra CD, Gluckman JC (eds) Accessory cells in HIV and other retroviral infections. Karger, Basel, pp 1–8

Heinen E, Tsunoda R, Bosseloir A et al. (1991b) Are germinal centers insulating microenvironments? In: Imhof BA, Berrih-Aknin S, Ezine S (eds) Lymphatic tissues and in vivo immune responses. Dekker, New York, pp 365–368

Heinen E, Tsunoda T, Marcoty C et al. (1991c) The germinal centre—a monastery or a bar. Res Immunol 142: 242–244

Heinen E, Tsunoda R, Marcoty C et al. (1993) Follicular dendritic cells: isolation procedures, short and long term cultures. Adv Exp Med Biol 329: 333–338

Heusermann UH, Zurborn UH, Schroeder L, Stutte HJ (1980) The origin of the follicular dendritic cell: an experimental enzyme-histochemical and electron microscopical study of the rabbit spleen. Cell Tissue Res 209: 279–294
Heyman B (1990) The immune complex: possible ways of regulating the antibody response. Immunol. Today 11: 310–313
Hodgkin PD, Castle BE, Kehry MR (1994) B cell differentiation induced by helper T cell membranes—evidence for sequential isotype switching and a requirement for lymphokines during proliferation. Eur J Immunol 24: 239–246
Hoefakker S, Vanterve EHM, Deen C et al. (1993) Immunohistochemical detection of co-localizing cytokine and antibody producing cells in the extrafollicular area of human palatine tonsils. Clin Exp Immunol 93: 223–228
Hoefsmit ECM, Kamperdjik EWA, Hendriks HR, Beelen RHJ, Balfour BM (1980) Lymph node macrophages. Reticuloendoth Syst 1: 417–468
Humphrey JH, Grennan D, Sundaram V (1984) The origin of follicular dendritic cells in the mouse and the mechanism of trapping of immune complexes on them. Eur J Immunol 14: 859–863
Imai Y, Kazuo T, Dobashi M, Maeda K, Kasajima T (1983) Reticulum cell and dendritic reticulum cell: origin and function. Recent Adv RES Res 21: 51–81
Imai Y, Yamakawa M, Masuda A, Sato T, Kasajima T (1986) Function of the follicular dendritic cell in the germinal center of lymphoid follicles. Histol Histopathol 1: 341–353
Imazeki N, Senoo A, Fuse Y (1992) Is the follicular dendritic cell a primarily stationary cell? Immunology 76: 508–510
Jacob J, Przylepa J, Miller C, Kelsoe G (1993) In situ studies of the primary immune response to (4-hydroxy-3-nitrophenyl) acetyl.3. The kinetics of V-region mutation and selection in germinal center B-cells. J Exp Med 178: 1293–1307
Jacobson EB, Caporale LH, Thorbeke GJ (1974) Effect of thymus cell injections on germinal center formation in lymphoid tissues of nude mice. Cell Immunol 13: 416–423
Johnson GD, Hardie DC, Ling NR, MacLennan ICM (1986) Human follicular dendritic cells (FDC): a study with monoclonal antibodies (MoAb). Clin Exp Immunol 64: 205–213
Kadmon G, Ekert M, Sammar M, Schachner M, Altevogt P (1992) Nectadrin, the heat-stable antigen, is a cell adhesion molecule. J Exp Med 118: 1245–1251
Kapasi ZF, Kosco MH, Schultz LD, Tew JG, Szakal AR (1994) Cellular origin of follicular dendritic cells. Adv Exp Med Biol (in press)
Kepler TB, Perelson AS (1993) Cyclic re-entry of germinal center B-cells and the efficiency of affinity maturation. Immunol Today 14: 412–415
Klaus GGB, Humphrey J (1977) The generation of memory cells. The role of C3 in the generation of B memory cells. Immunology 33: 31–40
Klaus GGB, Humphrey JH, Kunkel A, Dongworth DW (1980) The follicular dendritic cell: its role in antigen presentation in the generation of immunological memory. Immunol Rev 53: 3–28
Kölsch E, Oberbarnscheidt J, Brüner K, Heuer J (1983) The Fc-receptor: its role in the transmisssion of differentiation signals. Immunol Rev 49: 61–78
Koopman G, Pals ST (1992) Cellular interactions in the germinal center—role of adhesion receptors and significance for the pathogenesis of AIDS and malignant lymphoma. Immunol Rev 126: 21–45
Koopman G, Parmentier HK, Schuurman HJ, Newman W, Meijer CJLM, Pals ST (1991) Adhesion of human-B cells to follicular dendritic cells involves both the lymphocyte function-associated antigen-1 intercellular adhesion molecule-1 and very late antigen-4/vascular cell adhesion molecule-1 pathways. J Exp Med 173: 1297–1304
Korsmeyer SJ (1992) Bcl-2: a repressor of lymphocyte death. Immunol Today 13: 285–288
Kosco M, Szakal AK, Tew JG (1988) In vivo obtained antigen presented by germinal center B cells to T cells in vivo. J Immunol 140: 354–360
Kosco M, Burton GF, Kapasi ZF, Szakal AK, Tew JG (1989) Antibody-forming cell induction during an early phase of germinal center development and its delay with ageing. Immunology 68: 312–318
Kosco MH (1991a) Antigen presentation to B-cells. Curr Opin Immunol 3: 336–339
Kosco MH (1991b) Cellular interactions during the germinal centre response. Res Immunol 142: 245–248
Kosco MH, Gray D (1992) Signals involved in germinal center reactions. Immunol Rev 126: 63–76
Kosco MH, Pflugfelder E, Gray D (1992) Follicular dendritic cell-dependent adhesion and proliferation of B-cells invitro. J Immunol 148: 2331–2339
Kosco-Vilbois MH, Gray D, Scheidegger D, Julius M (1993) Follicular dendritic cells help resting B cells to become effective antigen-presenting cells: induction of B7/BB1 and upregulation of major histocompatibility complex class II molecules. J Exp Med 178: 2055–2066

Kromer G, Cuende E, Martinez AC (1993) Compartmentalization of the peripheral immune system. Adv Immunol 53: 157–216

Kroese FGM (1987) The generation of germinal centers. Doctoral thesis, Krips Repro Meppel, Groningen

Kroese FGM, Wubbena AS, Joling P, Nieuwenhuis P (1985) T lymphocytes in rat lymphoid follicles are a subset of T helper cells. Adv Exp Med Biol 186: 443–450

Kroese FGM, Wubbena AS, Nieuwenhuis P (1986) Germinal center formation and follicular antigen trapping in the spleen of lethally X-irradiated and reconstituted rats. Immunology 57: 99–104

Kroese FGM, Wubbena AS, Seijen H, Nieuwenhuis P (1987) Germinal centers develop oligoclonally. Eur J Immunol 17: 1069–1072

Kroese FGM, Seijen HG, Nieuwenhuis P (1991) The initiation of germinal centre reactivity. Res Immunol 142: 249–252

Kunkl A, Klaus GGB (1981) The generation of memory cells. IV. Immunization with antigen–antibody complexes accelerates the development of B memory cells, the formation of germinal centers and the maturation of antibody affinity in the secondary response. Immunology 43: 371–378

Kuppers R, Zhao M, Hansmann ML, Rajewsky K (1993) Tracing B-cell development in human germinal centres by molecular analysis of single cells picked from histological sections. EMBO J 12: 4955-4967

Lagresle C, Bella C, Defrance T (1993) Phenotypic and functional heterogeneity of the IgD-B-cell compartment—identification of two major tonsillar B-cell subsets. Int Immunol 5: 1259–1268

Lampert IA, Schofield JB, Amlot P, Vannoorden S (1993) Protection of germinal centres from complement attact—Decay-Accelerating Factor (DAF) is a constitutive protein on follicular dendritic cells—a study in reactive and neoplastic follicles. J Pathol 170: 115–120

Lane P, Burdet C, Hubele S et al. (1994) B cell formation in mice transgenic for mCTL-Hgamma1: lack of germinal centers correlated with poor affinity maturation and class switching despite normal priming of CD4+ T cells. J Exp Med 179: 819–830

Lebecque SG, Gearhart PJ (1990) Boundaries of somatic mutations in rearranged immunoglobulin genes. J Exp Med 172: 1717–1721

Lederman S, Yellin MJ, Inghirami G, Lee JJ, Knowles DM, Chess L (1992) Molecular interactions mediating lymphocyte-T-B collaboration in human follicles. Roles of T-cell-B-cell activating molecule (5c8 antigen) and CD40 in contact-dependent help. J Immunol 149: 3817–3828

Lennert K (1978) Malignant lymphomas other than Hodgkin's disease. Springer, Berlin Heidelberg New York

Lilet-Leclercq C, Radoux D, Heinen E et al. (1984) Isolation of follicular dendritic cells from human tonsils and adenoids. I. Procedure and morphological characterization. J Immunol Methods 66: 235-244

Lindhout E, Mevissen MLCM, Kwekkeboom J, Tager JM, Degroot C (1993) Direct evidence that human Follicular Dendritic Cells (FDC) rescue germinal centre B-cells from death by apoptosis. Clin Exp Immunol 91: 330–336

Linton PJ, Rudie A, Klinman NR (1991) Tolerance susceptibility of newly generating memory B-cells. J Immunol 146: 4099–4104

Liu YJ, Joshua DE, Williams GT, Smith CA, Gordon J, MacLennan ICM (1989) Mechanism of antigen-driven selection in germinal centers. Nature 342: 929–931

Liu YJ, Cairns JA, Holder MJ et al. (1991a) Recombinant 25-kDa CD23 and interleukin-1 alpha promote the survival of germinal center B-cells—evidence for bifurcation in the development of centrocytes rescued from apoptosis. Eur J Immunol 21: 1107–1114

Liu YJ, Mason DY, Johnson GD et al. (1991b) Germinal center cells express bcl–2 protein after activation by signals which prevent their entry into apoptosis. Eur J Immunol 21: 1905–1910

Louis E, Philippet B, Cardos B et al. (1989) Intercellular connections between germinal center cells. Mechanisms of adhesion between lymphoid cells and follicular dendritic cells. Acta Otolaryngol (Stockh) 43: 297–32

Mackey CR (1993) Immunological memory. Adv Immunol 53: 217–265

Manser T (1989) Evolution of antibody structure during the immune response. J Exp Med 170: 1211–1230

Manser T (1990) The efficiency of antibody affinity maturation: can the rate of B-cell division be limiting? Immunol Today 11: 305–308

Manser T, Wysocki L, Margolies MN, Gefter ML (1987) Evolution of antibody variable region structure during the immune response. Immunol Rev 96: 141–159

Mason DY, Jones M, Goodnow CC (1992) Development and follicular localization of tolerant lymphocytes-B in lysozyme/ anti-lysozyme IgM/IgD transgenic mice. Int Immunol 4: 163–175

Mattheys JB, Basu MK (1982) Oral tonsils: an immunoperoxidase study. Int Arch Allergy Appl Immunol 69: 21–25

McDonnell TJ, Korsmeyer SJ (1991) Progression from lymphoid hyperplasia to high-grade malignant mice transgenic for the t(14;18). Nature 349: 254–256
Miyawaki T, Uehara T, Nibu R et al. (1992) Differential expression of apoptosis-related fas antigen on lymphocyte subpopulations in human peripheral blood. J Immunol 149: 3753-3758
Mori S, Mohri N, Morita H, Yamaguchi K, Shimamine T (1983) The distribution of cells expressing a natural Killer marker (HNK1) in normal human lymphoid and malignant lymphomas. Virchows Arch 43: 253–263
Munro JM, Freedman AS, Aster JC et al (1994) In vivo expression of the B7 costimulatory molecule by subsets of antigen–presenting cells and the malignant cells of Hodgkin's disease. Blood 83: 793–798
Muta T, Kurosaki T, Misulovin Z, Sanchez M, Nussenzweig MC, Ravetch JV (1994) A 13-amino acid motif in the cytoplasmic domain of FCgammaRIIB modulates B-cell receptor signalling. Nature 368: 70–72
Naiem M, Gerdes J, Abdulaziz Z, Stein H, Mason DY (1983) Production of a monoclonal antibody reactive with human dendritic reticulum cells and its use in the immunohistological analysis of lymphoid tissue. J Clin Pathol 36: 167–175
Namikawa R, Mizuno T, Matsuoka H et al. (1986) Ontogenic development of T cells and B cells and non-lymphoid cells in the white pulp of human spleen. Immunology 57: 61–70
Newell MK, Vanderwall J, Beard K, Freed JH (1993) Ligation of major histocompatibility complex class II molecules mediates apoptotic cell death in resting B lymphocytes. Proc Natl Acad Sci USA 90: 10459–10463
Nossal GJV (1994) Negative selection of lymphocytes. 76: 229–239
Nossal GJV, Ada GL, Austin CM, Pye J (1965) Antigens in immunity. VIII. Localization of 125-I-labeled antigens in the secondary response. Immunology 9: 349–356
Nunez G, Hockenbery D, McDonnell TJ, Sorensen CM, Korsmeyer SJ (1991) Bcl-2 maintains B-cell memory. Nature 353: 71–73
Okada M, Sakaguchi N, Yohimura N, Hara H, Shimizu K, Yoshida N, Yashizahi K, Kishimoto S, Yamamura Y, Kishimoto T (1983) B cell growth factors and B cell differentiation factor from human T hybridomas. Two distinct kinds of B cell growth factors and their synergism in B cell proliferation. J Exp Med 157: 583–590
Oltvai ZN, Milliman CL, Korsmeyer SJ (1993) Bcl-2 heterodimerizes in vivo with conserved homology, Bax, that accelerates programed cell death. Cell 74: 609–619
Opstelten D, Deenen GJ, Bos J, Nieuwenhuis P (1982) Localization of germinal center cell subsets differing in density and sedimentation velocity. Adv Exp Med Biol 149: 757–764
Pallesen G, Myrhe-Jensen O (1987) Immunophenotypic analysis of neoplastic cells in follicular cell sarcoma. Leukemia 1: 549–557
Papamichail M, Gutierrez C, Embling P, Johnson P, Holborrow EJ, Peys MB (1975) Complement dependency of localization of aggregated IgG in germinal centers. Scand J Immunol 4: 343–347
Parwaresch MR, Radzun HJ, Hansman ML, Peters KP (1983) Peroxidase-positive mononuclear precursors of human dendritic reticular cells. J Immunol 131: 2719–2725
Petrasch SG, Stein H, Kosco MH, Brittinger G (1991a) Follicular dendritic cells in non-Hodgkin lymphomas—localisation, characterisation and pathophysiological aspects. Eur J Cancer 27: 1052–1056
Petrasch SG, Kosco MH, Perezalvarez CJ, Schmitz J, Brittinger G (1991b) Proliferation of germinal center B-lymphocytes in vitro by direct membrane contact with follicular dendritic cells. Immunobiology 183: 451–462
Phillips NE, Parker DC (1985) Subclass specificity of Fcgamma receptor mediated inhibition of mouse B cell activation. J Immunol 134: 2835–2838
Phipps RP, Mandel TE, Schnizlein CT, Tew JG (1984) Anamnestic responses induced by antigen persisting on follicular dendritic cells from cyclophosphamide-treated mice. Immunology 51: 387–397
Pizzolo G, Semenzato G, Chilosi M et al. (1984) Distribution and heterogeneity of cells detected by HNK-1 monoclonal antibody in blood and tissues in normal reactive and neoplastic conditions. Clin Exp Immunol 57: 195–206
Poppema S, Bhan AK, Reinherz EC, McCluskey RT, Schlossman SF (1989) Distribution of T cell subsets in human lymph nodes. Eur J Immunol 19: 481–485
Porwit-Ksiazek A, Ksiazek T, Biberfeld P (1983) Leu-7$^+$ cells. I. Selective compartmentalization of LEU-7 cells with different immunophenotypes in lymphatic tissues and blood. Scand J Immunol 18: 485–493

Rademakers LHPM (1992) Dark and light zones of germinal centres of the human tonsil—an ultrastructural study with emphasis on heterogeneity of follicular dendritic cells. Cell Tissue Res 269: 359–368

Rademakers LHPM, De Weger RA, Roholl PJM (1989) Identification of alkaline phosphatase positive cells in human germinal centers as follicular dendritic cells. Adv Exp Med Biol 237: 165–168

Radoux D, Heinen E, Kinet-Denoël C, Tihange E, Simar LJ (1984) Precise localization of antigens on follicular dendritic cells. Cell Tissue Res 235: 257–274

Radoux D, Heinen E, Kinet-Denoël C, Simar LJ (1985a) Antigen–antibody retention by follicular dendritic cells. Adv Exp Med Biol 186: 185–192

Radoux D, Kinet-Denoël C, Heinen E, De Mey J, Moeremans M, Simar LJ (1985b) Retention of immune complexes by Fc receptors on mouse follicular dendritic cells. Scand J Immunol 21: 345–352

Rice GE, Munro JM, Corless C, Bevilacqua MP (1991) Vascular and nonvascular expression of INCAM-110. A target for mononuclear leukocyte adhesion in normal and inflamed human tissues. Am J Pathol 138: 385–393

Rigley KP, Harnett MM, Klaus GGB (1989) Co-crosslinking of surface immunoglobulin Fcgamma receptors on B lymphocytes uncouples the antigen receptors from their associated G protein. Eur J Immunol 19: 481–485

Ritchie AWS, James K, Micklem HS (1983) The distribution and possible significance of cells identified in human lymphoid tissue by the monoclonal antibody HNK-1. Clin Exp Immunol 51: 439–447

Rodriguez C, Bellas C, Brieva JA (1992) Human germinal centre B-cells inhibit mitogen-induced proliferation of mantle zone B-cells. Scand J Immunol 35: 745–750

Roes J, Rajewsky K (1993) Immunoglobulin D (IgD-deficient) mice reveal an auxilliary receptor function for IgD in antigen-mediated recruitment of B cells. J Exp Med 177: 45–55

Rozing J, Brons NHC, VanEwijk W, Benner R (1978) B lymphocyte differentiation in lethally irradiated and reconstituted mice: a histological study using immunofluorescent detection of B lymphocytes. Cell Tissue Res 189: 19–30

Ruco LP, Gearing AJH, Pigott R et al. (1991) Expression of ICAM-1, VCAM-1 and ELAM-1 in angiofollicular lymph node hyperplasia (Castleman's disease)—evidence for dysplasia of follicular dendritic reticulum cells. Histopathology 19: 523–528

Sablitzky F, Wildner G, Rajewsky K (1985) Somatic mutation and clonal expansion of B cells in an antigen driven immune response. EMBO J 4: 345–350

Sandor M, Lynch R (1993) Lymphocyte Fc receptors: the typical case of T cells. Immunol Today 14: 227–231

Schmitz J, Petrasch S, Vanlunzen J et al. (1993) Optimizing follicular dendritic cell isolation by discontinuous gradient centrifugation and use of the Magnetic Cell Sorter (MACS). J Immunol Methods 159: 189–196

Schriever F, Nadler LM (1991) Antigenic phenotype of isolated human follicular dendritic cells. In: Racz P, Dijkstra CD, Gluckman JC (eds) Accessory cells in HIV and other retroviral infections. Karger, Basel, pp 9–17

Schriever F, Nadler LM (1992) The central role of follicular dendritic cells in lymphoid tissues. Adv Immunol 51: 242–284

Schriever F, Freedman AS, Freeman G et al. (1989) Isolated human follicular dendritic cells display a unique antigenic profile. J Exp Med 169: 2043–2058

Seijen HG, Bun JCA, Wubbena AS, Löhlefink KJ (1988) The germinal center precursor cell is surface μ and δ positive. Adv Exp Med Biol 237: 233–237

Si, Whiteside TL (1983) Tissue distribution of human NK cells studied with anti-Leu7 monoclonal antibody. J Immunol 130: 2149–2155

Simar LJ, Weibel ER (1979) Plasma cell differentiation in lymphoid tissue. In: Weibel ER (ed) Sterological methods. Academic, London, pp 337–342

Sinclair NR (1969) Regulation of the immune response. I. Reduction in ability of specific antibody to inhibit long lasting IgG immunological priming after removal of the Fc fragment. J Exp Med 129: 1183–1191

Siskind GW, Benacerraf B (1969) Cell selection by antigen in the immune response. Adv Immunol 10: 1–29

Stein H, Gerdes J, Mason DY (1982) The normal and malignant germinal centre. Clin Haematol 11: 531–559

Strasser A, Whittingham S, Vaux DL et al. (1990) Enforced BCL2 expression in B-lymphoid cells prolongs antibody responses and elicits autoimmune disease. Proc Natl Acad Sci USA 88: 8661–8665

Szakal AK, Kosco M, Tew JG (1988) A novel in vivo follicular dendritic cell-dependent iccosome-mediated mechanism for delivery of antigen to antigen-processing cells. J Immunol 140: 341–353
Szakal AK, Kosco M, Tew JG (1989) Microanatomy of lymphoid tissue during humoral immune responses. Annu Rev Immunol 7: 91–109
Terashima K, Dobashi M, Maeda K, Imai Y (1991) Cellular components involved in the germinal centre reaction. Res Immunol 142: 263–268
Tew JG, Mandel TE, Burgess AW (1979) Retention of intact HSA for prolonged periods in the popliteal lymph nodes of specifically immunized mice. Cell Immunol 45: 207–212
Tew JG, Phipps RP, Mandel TE (1980) The maintenance and regulation of the humoral immune response: persisting antigen and the role of follicular antigen-binding cells as accessory cells. Immunol Rev 53: 175–201
Thorbeke GJ, Asofsky RM, Hochwald GM, Siskind GM (1962) Gamma-gobulin and antibody formation in vitro. III. Induction of secondary response at different intervals after the primary; the role of secondary nodules in the preparation of the secondary response. J Exp Med 116: 295–310
Trauth BC, Klas C, Peters AMJ, Matzku S, Möller P, Falk W, Debatin KM, Krammer PH (1989) Monoclonal antibody-mediated tumor regression by induction of apoptosis. Science 245: 301–305
Tsiagbe VK, Linton PJ, Thorbecke GJ (1992) The path of memory B-cell development. Immunol Rev 126: 113–141
Tsubata T, Wu J, Honjo T (1993) B-cell apoptosis induced by antigen receptor crosslinking is blocked by a T-cell signal through CD 40. Nature 364: 645–648
Tsunoda R, Yaginuma Y, Kojima M (1980) Immunological studies on the constituent cells of secondary nodules in human tonsils. Acta Pathol Jpn 30: 33–57
Tsunoda R, Kojima M (1982) Immunocytological characterization of the constituent cells of the secondary nodules in human tonsils. Adv Exp Med Biol 149: 829–834
Tsunoda R, Cormann N, Heinen E et al. (1989) Cytokines produced in lymph follicles. Immunol Lett 22: 129–134
Tsunoda R, Nakayama M, Onozaki K et al. (1990) Isolation and longterm cultivation of tonsil follicular dendritic cells. Virchows Arch [B] 59: 95–105
Tsunoda R, Nakayama M, Heinen E et al. (1992) Emperipolesis of lymphoid cells by human follicular dendritic cells in vitro. Virchows Arch [B] 62: 69–78
Urbain J (1986) Idiotypic networks: a noisy background or a breakthrough in immunological thinking. Ann Inst Pasteur Immunol 135: 57–64
Van Noesel CJM, Lankester AC, vanLier RAN (1993) Dual antigen recognition by B cells. Immunol Today 14: 8–11
Van Rooijen N (1980) Immune complex trapping in lymphoid organs: a discussion on possible functional implications. In: Manning MJ (ed) Phylogeny of immunological memory. Elsevier, Amsterdam, pp 281–290
Van Rooijen N (1991) The role of antigens, antibodies and immune complexes in the functional activity of germinal centres. Res Immunol 142: 272–275
Velardi A, Mingari MC, Moretta L, Grossi CE (1986a) Functional analysis of cloned germinal center CD4+ cells with natural killer cell-related features. Divergence from typical T helper cells. J Immunol 137: 2808–2813
Velardi A, Tilden AB, Millo R, Grossi CE (1986b) Isolation and characterization of Leu7+ germinal-center cells with the helper cell phenotype and granular lymphocyte morphology. J Clin Immunol 6 205–215
Villena A, Zapata A, Rivera-Pomar JM, Barrutia MJ, Fonfria J (1983) Structure of the non-lymphoid cells during the postnatal development of the rat lymph nodes. Fibroblastic reticulum cells and interdigitating cells. Cell Tissue Res 229: 219–232
Vonderheide RH, Hunt SV (1990) Does availability of either B cells or CD4+ cells limit germinal centre formation? Immunology 69: 487–489
Wacker HH, Heidelbrecht HJ, Radzun HJ, Parwaresch MR (1991) Dendritic cells: morphofuctional analysis in health and disease. In: Imai Y, Tew JC, Hoefsmit ECM (eds) Dendritic cells in lymphoid tissues. Elsevier, New York, pp 187–191
Wang DN, Wells SM, Stall AM, Kabat EA (1994) Reaction of germinal centers in the T-cell-independent response to the bacterial polysaccharide-alpha (1–6) dextran. Proc Natl Acad Sci USA 91: 2502–2506
Weiss U, Zoebelein R, Rajewsky K (1992) Accumulation of somatic mutants in the B-cell compartment after primary immunization with a T-cell-dependent antigen. Eur J Immunol 22: 511–517
Williams GM, Nossal GJV (1966) Ontogeny of the immune response: I. The development of the follicular antigen-trapping mechanism. J Exp Med 124: 47–56

Wyllie AH, Morris RG, Smith AL, Dunlop D (1984) Chromatin cleavage in apoptosis: association with condensed chromatin morphology and dependence on macromolecular synthesis. J Pathol 142: 67–78

Yamakawa M, Imai Y (1992) Complement activation in the follicular light zone of human lymphoid tissues. Immunology 76: 378–384

Yoshida K, Vandenberg TK, Dijkstra CD (1993) Two functionally different follicular dendritic cells in secondary lymphoid follicles of mouse spleen, as revealed by CR1/2 and FcR-gamma-II-mediated immune-complex trapping. Immunology 80: 34–39

Yoshida K, Vandenberg TK, Dijkstra CD (1994) The functional state of follicular dendritic cells in Severe Combined Immunodeficient (SCID) mice—role of the lymphocytes. Eur J Immunol 24: 464–468

Mechanism of Immune Complex Trapping by Follicular Dendritic Cells

T.K. VAN DEN BERG[1], K. YOSHIDA[2], and C.D. DIJKSTRA[1]

1 Introduction

The follicular dendritic cell (FDC) is a major constituent of the microenvironment of the lymphoid follicle. FDC have the characteristic and unique property of binding antigens in the form of antigen–antibody complexes and retaining these complexes, without ingestion, for long periods of time. The presence of antigen–antibody complexes, also called immune complexes, on FDC is believed to play a crucial role in the development of B cell memory and the affinity maturation of the antibody response against T cell-dependent antigens.

In order to understand the role of immune complexes trapped by FDC in the establishment of the immune response, we have to understand the mechanism

[1] Department of Cell Biology and Immunology, Vrije University, Van Der Boechorststraat 7, 1081 BT Amsterdam, The Netherlands
[2] Department of Molecular Immunology, Institute of Development, Aging and Cancer, Tohoku University, 4-1 Seiryomachi, Aoka-ku, Sendai 980, Japan

of follicular trapping. This chapter focuses on the molecular and cellular requirements of the trapping process and the role of immune complexes on FDC in the activation of B lymphocytes and the generation of memory.

2 Immune Complex Trapping

It has been known for a long time that antigen can persist in lymphoid tissues. Early studies demonstrated that this retention only occurred in preimmunized animals (Humphrey and Franck 1967), suggesting that specific antibody was required and antigen was trapped in the form of antigen–antibody immune complexes. It was also shown that immune complexes were trapped and retained on the surface of specialized cells located in lymphoid follicles, and these were called FDC (Nossal and Ada 1968). The trapping of immune complexes by FDC in vivo has been studied by injection of antigen in actively and passively immunized animals and by injection of preformed immune complexes or aggregated immunoglobulins. Most of our knowledge on the trapping of immune complexes is derived from studies in rodents, but the phenomenon is also observed in birds (White et al. 1975; Eikelenboom et al. 1983), fish (Secombes et al. 1982), amphibians (Collie 1974), and reptiles (Kroese and Van Rooijen 1983; Kroese et al. 1985).

2.1 Immune Complexes

Once an antigen is complexed to an antibody, it is called an immune complex. Therefore, if we are dealing with immune complexes in vivo, we are dealing with the effector functions of antibodies. These include: (a) activation of complement by either the alternative or classical pathway and (b) binding to Fc receptors.

The main consequence of complement activation is that C3 fragments (i.e., C3b, C3bi, and/or C3d) become deposited on the immune complex. Therefore, in the body an immune complex is a combination of antigen, antibody, and complement. It should be noted that the route (i.e., classical or alternative pathway) and extent of complement activation are strongly dependent on the class or isotype of the antibody, and this will affect the amount of C3 fragments present in the immune complex. Thus, immunoglobulin (Ig)M and IgA are poor activators of complement, whereas most IgG isotypes are good complement activators (Klaus et al. 1979). This is highly relevant, since the ability of an immune complex to activate complement strongly correlates with its trapping by FDC and its ability to generate B cell memory (Embling et al. 1978; Klaus 1979; Phipps et al. 1980). As a matter of fact, the presence of complement in an immune complex is the most important and in many circumstances the only requirement for trapping by

FDC. This was initially shown using in vivo treatment with cobra venom factor (CoVF), which strongly inhibits, and often completely prevents, follicular trapping in mice (PAPAMICHAIL et al. 1975; KLAUS and HUMPHREY 1977; CHEN et al. 1978; TEW et al. 1979; PHIPPS et al. 1980; ENRIQUEZ-RINCON et al. 1984), rats (GRAY et al. 1984b; VAN DEN BERG et al. 1992), and chickens (WHITE et al. 1975). CoVF is a powerful activator of the alternative pathway of complement and thereby causes C3 depletion from the circulation. In addition, we have shown that in vivo administration of a monoclonal antibody against C3, which prevents complement activation without affecting serum levels of C3 (or other complement factors), also abolishes trapping (Fig. 1; VAN DEN BERG et al. 1992). The fact that complement can be the only requirement for follicular trapping is illustrated by the observation that the bacterial antigen levan and nonimmunogenic compounds such as colloidal thorium or carbon particles, all of which can directly activate complement in the absence of antibody, rapidly localize in follicles upon injection (KLAUS et al. 1980; GROENEVELD et al. 1983). The role of complement and complement receptors will be described in more detail in Sect. 2.3.1.

The binding of immune complexes to Fc receptors is also dependent on the class or isotype of the antibody. For instance, IgM, which predominates during the early stages of a primary immune response, is unable to bind to Fc receptors. Thus, the only way for an IgM immune complex to bind to a cell is via complement activation and interaction with complement receptors on the cell surface. With

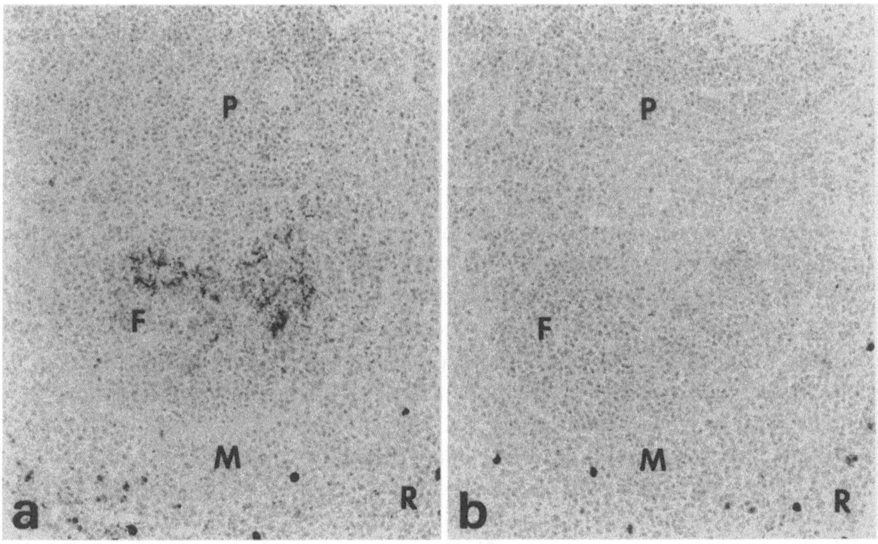

Fig. 1a, b. The role of C3 in the trapping of immune complexes in vivo. Normal rats were treated with control monoclonal antibody (mAb) (**a**) or anti-C3 mAb ED11 (**b**), followed 2 h later by i.v. injection of rabbit peroxidase–anti-peroxidase complexes. Trapping was analyzed in the spleen 24 h after the injection of immune complexes. Anti-C3 treatment completely prevented trapping *F*, follicle; *P*, periarteriolar lymphocyte sheath; *M*, marginal zone; *R*, red pulp. (Reproduced from VAN DEN BERG et al. 1992, with permission; x 200)

respect to follicular trapping, the Fc part of the antibody in an immune complex is important, but not crucial. The trapping of immune complexes with $F(ab')_2$ fragments can occur, but is much less than with intact antibody (KLAUS et al. 1980). Whether reduced binding is due to loss of complement activation by the classical pathway ($F(ab')_2$ immune complexes are still able to activate complement by the alternative route), loss of the ability to bind to Fc receptors, or both, is not known. As will be discussed in Sect. 2.3.2, it seems probable that under particular circumstances Fc-mediated trapping does indeed occur.

When considering the trapping of immune complexes by FDC, it is important to bear in mind that in vivo most (> 99%) immune complexes are rapidly ingested and degraded by phagocytic cells (i.e., macrophages and granulocytes). Kupffer cells in the liver take up most immune complexes from the circulation. In the spleen and lymph nodes, most immune complexes are ingested by macrophages of the marginal zone and marginal sinus, respectively.

2.2 Transport of Immune Complexes to the Follicular Dendritic Cells

In order to reach FDC located in lymphoid follicles, immune complexes have to migrate from either the blood or from the afferent lymph to the spleen and draining lymph nodes, respectively. There has been much discussion as to whether this transport is an active process, mediated by transporting cells that bind immune complexes and carry them to the follicles, or whether it is a passive process, that occurs by simple diffusion of immune complexes through the interstitial tissue fluid.

2.2.1 Spleen

Immune complexes from the circulation enter the spleen in the marginal zone, where they become associated with macrophages and marginal zone B cells. Most immune complexes are phagocytosed and degraded by macrophages, but some migrate to the lymphocyte corona and subsequently to the follicle, where they are trapped by FDC. Based on experiments with nonimmunogenic carbon particles, GROENEVELD et al. (1983) concluded that transport (and trapping) occurs by diffusion. However, since carbon particles are able to fix complement by the alternative pathway, complement-mediated transport cannot be excluded. It also seems clear that macrophages are not involved in the transport of immune complexes to the follicles, as macrophage depletion in vivo does not prevent follicular trapping (LAMAN et al. 1990). On this point we favor the hypothesis that B lymphocytes mediate the transport of immune complexes to the FDC for the following reasons:

1. B cells preincubated with immune complexes are able to deliver them to the FDC in vivo (BROWN et al. 1970) or in vitro (HEINEN et al. 1986a).

2. Trapping is prevented by depletion of total lymphocytes using X-irradiation (KROESE et al. 1986; VAN DEN BERG et al. 1992), but can still occur in the absence of T cells (KLAUS and HUMPHREY 1977; MJAALAND and FOSSUM 1987).
3. Trapping is prevented by selective depletion of marginal zone B cells by cyclophosphamide (GRAY et al. 1984a).

It should be noted that the role of marginal zone B cells in immune complex transport can also in part explain the effects of lipopolysaccharide (LPS) on follicular trapping. LPS administration induces massive migration of marginal zone B cells to the follicles, and this results in the depletion of B cells from the marginal zone (GRAY et al. 1984b). Immune complexes injected early after LPS show increased localization in the follicles, whereas trapping is inhibited after B cells have been depleted (VAN ROOIJEN et al. 1975; GRAY et al. 1984b). On the other hand, LPS can also affect the FDC itself, as it results in the removal of immune complexes already trapped on FDC (GROENEVELD et al. 1983; VAN DEN BERG et al. 1992) and inhibits the binding of immune complexes to FDC in vitro (DIJKSTRA et al. 1983; HEINEN et al. 1986b). Whether this is due to downregulation of receptors for immune complexes is not known.

2.2.2 Lymph Nodes

Compared to the situation in the spleen, little is known about the mechanism of immune complex transport in lymph nodes. Immune complexes enter the lymph nodes via the afferent lymphatics, and the transport of immune complexes to the follicles has been described (KAMPERDIJK et al. 1987). Initially, immune complexes can be found associated with several different cell types, including macrophages and lymphocytes, but there is no direct evidence from in vivo experiments that these cells are involved in the transport of immune complexes to the FDC. SZAKAL et al. (1983, 1993) have shown that part of the antigen or immune complexes migrating to the follicles is associated with so-called antigen-transporting cells. These cells have morphological and phenotypical characteristics of macrophages and dendritic cells, and it has been suggested that they can transform into FDC. Whether these cells are related to the so-called Langerhans cells, which are known to transport antigens to the T-dependent areas of the lymphoid tissue, and whether they are indeed responsible for the transport of immune complexes to the FDC is not clear at present.

2.3 Binding of Immune Complexes to the Follicular Dendritic Cells

The interaction between immune complexes and FDC has been studied in several ways. In spite of the fact that FDC constitute only a small percentage of the cells in lymphoid tissues, HEINEN et al. (1985) and KOSCO et al. (1986) have developed protocols for the isolation of these cells from human and mouse tissues, respectively, and have studied their immune complex-binding properties.

The main disadvantages of studying isolated FDC are that the possibility that the isolated cells only constitute a subpopulation of FDC can never be excluded and FDC are known to be heterogeneous. Furthermore, FDC can only be isolated from stimulated animals, and these may have different properties than FDC from unstimulated animals. We have studied the binding of immune complexes to FDC in frozen sections. Disadvantages of the frozen section technique are: (a) it is not possible to discriminate between cell surface and intracellular binding and (b) the cells have lost their metabolic activity. In this section, the role of the different receptors that have been implicated in the binding of immune complexes to the FDC will be discussed (summarized in Table 1).

2.3.1 Role of Complement and Complement Receptors

As described in Sect 2.1, the process of follicular trapping is strictly complement dependent. This raised the possibility that immune complexes are trapped via C3 receptors located on FDC. In accordance with this, C3 fragments have been found in association with immune complexes on FDC (Fig. 2; Gajl-Peczalaska et al. 1969; Imai et al. 1986; Halstensen et al. 1988; Van den Berg et al. 1989, 1992), and binding of opsonized sheep erythrocytes to B cell-depleted follicles (Gray et al. 1984a) provided indirect evidence for the presence of C3 receptors on FDC.

Several receptors for C3 fragments have been described, including complement receptor 1 (CR1, CD35), complement receptor 2 (CR2, CD21), and complement receptor 3 (CR3, CD11b/CD18). The human CR1 and CR2 are composed of a number of short consensus repeat (SCR) units (30 and 15 or 16, respectively) characteristic of the regulation of complement activation (RCA) gene cluster (Fearon and Ahearn 1989). CR1 binds the complement components C3b and C4b, whereas CR2 recognizes C3d and the Epstein-Barr Virus glycoprotein gp350/220.

Table 1. Receptors implicated in binding of immune complexes by follicular dendritic cells (FDC)

Receptor	Ligands	Expression[a]	Functional[b]
Complement receptors			
CR1 (CD35)	C3b	+	+
CR2 (CD21)	C3d, EBV, IFN-α, CD23	+	+
CR3 (CD11b/CD18)	C3bi, ICAM-1, etc.	+/–(human)	n.t.
		– (mouse, rat)	–
Fc receptors			
FcγRI (CD64)	Monomeric IgG	–	–
huFcγRII (CDw32)	$IgG_1 > IgG_2 = IgG_4 > IgG_3$	+/–	n.t.
muFcγRII	IgG_1, IgG_{2a}, IgG_{2b}	+subpop[c]	+
FcγRIII (CD16)	IgG_1, IgG_3 (human)	+/-	n.t.
FcεRII (CD23)			IgE
		+subpop[c]	+

For references see text.
n.t., not tested; +/–, controversial; EBV, Epstein-Barr Virus; IFN, interferon; ICAM, intercellular adhesion molecule; Ig, immunoglobulin.
[a]Expression means that the indicated molecule has been detected (+) or not (–) on FDC by immunocytochemistry.
[b]Functional means that the indicated molecule is involved (+) or not (–) in immune complex binding, as indicated by antibody blocking studies.
[c]Only on a subpopulation of FDC located in the apical light zone of the secondary follicle.

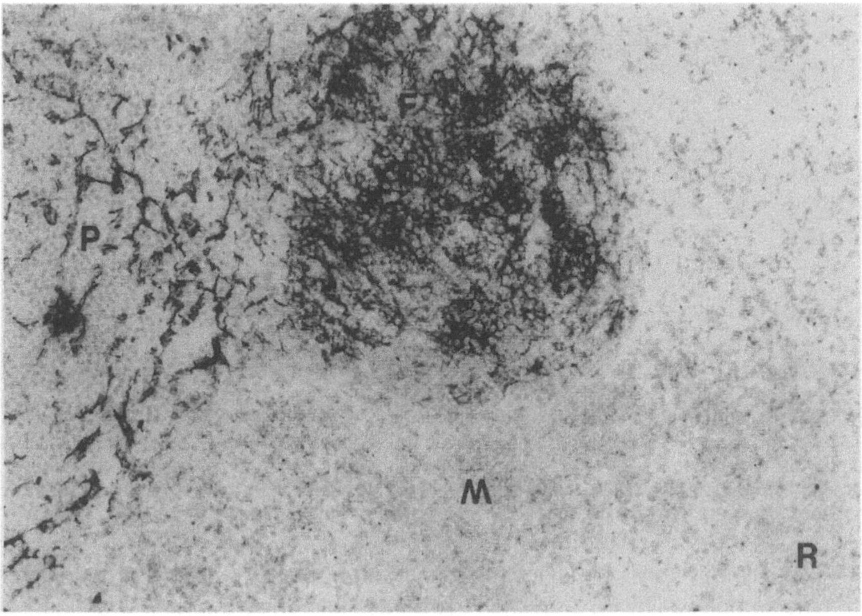

Fig. 2. Localization of C3 on follicular dendritic cells (FDC) in the rat spleen. Normal rat spleen stained with anti-C3 monoclonal antibody (mAb) ED11, which demonstrates the presence of C3 on FDC in the follicles (*F*) and on reticular elements in the outer periarteriolar lymphocyte sheath (*P*). *M*, marginal zone; *R*, red pulp. (Reproduced from VAN DEN BERG et al. 1992, with permission; x 100)

It has been established that FDC in human lymphoid tissues express high levels of CR1 and CR2 (REYNES et al. 1985), and this has been confirmed using isolated cells (SCHRIEVER et al. 1989; SELLHEYER et al. 1989; PETRASCH et al. 1990). However, to date there is no direct evidence for the involvement of CR1 and CR2 in the binding of immune complexes by human FDC.

In contrast to human CR1 and CR2, which are encoded by different genes, mouse CR1 and CR2 arise by alternative splicing from a gene that is most closely related to the human CR2 gene (FINGEROTH et al. 1989; FINGEROTH 1990; KURTZ et al. 1989; HOLERS et al. 1992). Compared to murine CR2, which is composed of 15 SCR and binds C3d (> C3b), murine CR1 contains an additional stretch of six SCR that exhibit C3b (> C3d) binding (KINOSHITA et al. 1988, 1990; MOLINA et al. 1992; PRAMOONJAGO et al. 1993). Recently, we have shown that binding of immune complexes to FDC in unstimulated animals can only occur in the presence of complement (Fig. 3). A monoclonal antibody (mAb) (7G6) against a common epitope on mouse CR1 and CR2 which blocks C3d binding is able to completely inhibit the binding of C3-coated immune complexes to FDC in the mouse spleen. An mAb against CR1 (8C12) which blocks C3b binding was partly inhibitory (YOSHIDA et al. 1993). This shows that CR1 and CR2 are indeed responsible for the binding of immune complexes to FDC, at least in normal animals.

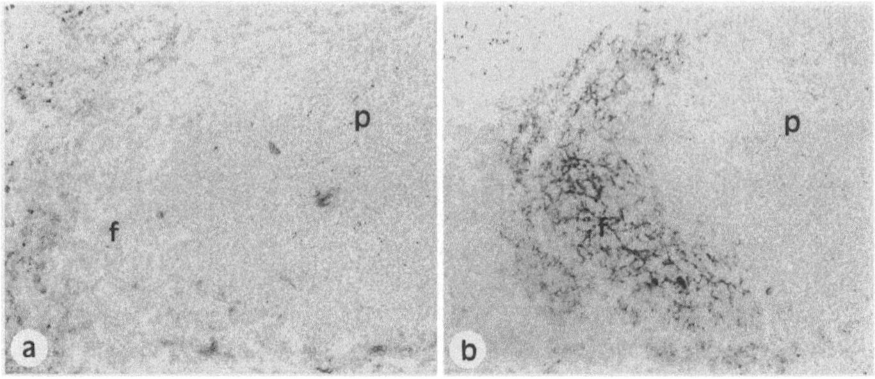

Fig. 3a, b. Binding of immune complexes to follicular dendritic cells (FDC) in the normal mouse spleen. Serial spleen sections were incubated with mouse immunoglobulin G (IgG) peroxidase–anti-peroxidase complexes in the absence (**a**) or presence (**b**) of fresh mouse serum as a source of complement. Immune complex binding only occurs in the presence of complement. *f*, follicle; *P*, periarteriolar lymphocyte sheath. (Reproduced from YOSHIDA et al. 1993, with permission; x 100)

There is only very little evidence to support a role for CR3 in the trapping of immune complexes by FDC. CR3, in the mouse known as the MAC-1 antigen, is an integrin composed of the α_m-chain (CD11b) and the noncovalently associated β_2-chain (CD18). It can bind to C3bi, intercellular adhesion molecule (ICAM)-1, and extracellular ligands (ROSEN and LAW 1989). The expression of CD11b on human FDC appears to be either weak (GERDES et al. 1983; REYNES et al. 1985; SCHRIEVER et al. 1989; PETRASCH et al. 1990) or absent (IMAI et al. 1986), whereas CD18 is not detected (SCHRIEVER et al. 1989). Studies in mouse (KOSCO et al. 1986; YOSHIDA et al. 1993) and rat (DAMOISEAUX et al. 1989; GRAY et al. 1984) indicate that CR3 is not expressed by FDC in these species.

2.3.2 Role of Fc Receptors

Several different types of receptors for IgG (FcγR) have been characterized, including FcγRI (CD64), FcγRII (CDw32), and FcγRIII (CD16). FcγRI is the high-affinity receptor for IgG and the only one that can bind monomeric IgG. This receptor is apparently absent from FDC (SELLHEYER et al. 1989). The expression of FcγRIII is controversial. One group has detected FcγRIII on isolated human FDC (PETRASCH et al. 1990), although they did not provide evidence that it was functional, whereas others have claimed that FDC are FcγRIII negative (SCHRIEVER et al. 1989; SELLHEYER et al. 1989).

We have recently characterized the role of FcγRII in the binding of immune complexes to FDC in the mouse using a frozen section assay (Fig. 4; YOSHIDA et al. 1993). Our results showed that, due to the absence of FcγRII from FDC in the spleen of normal (unimmunized) animals, there is no detectable Fc-mediated binding of immune complexes. However, antigenic stimulation induced the expression of FcγRII on a subpopulation of FDC located in the light zone of the secondary follicle and these FDC were able to bind immune complexes in

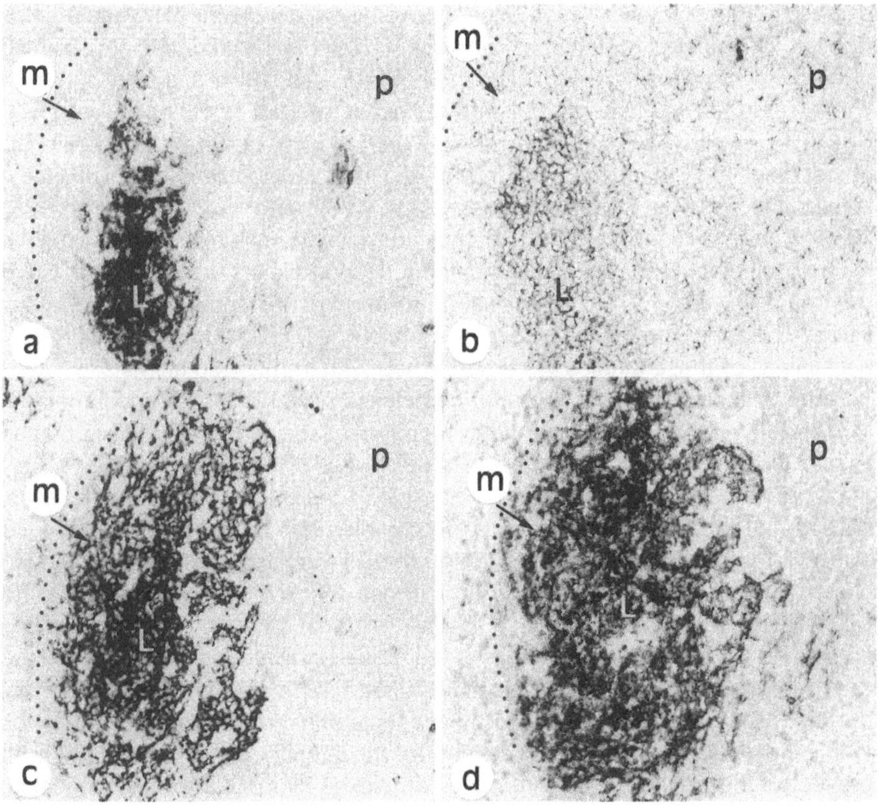

Fig. 4a–d. Binding of immune complexes to follicular dendritic cells (FDC) in the spleen from sheep red blood cell (SRBC)-stimulated mouse in relation to the expression of Fcγ receptor II (FcγRII) and complement receptors. Serial spleen sections were incubated with mouse immunoglobulin G (IgG) peroxidase–anti-peroxidase complexes in the absence (**a**) or presence of fresh mouse serum (**c**) as a source of complement. The distribution of FcγRII (monoclonal antibody, mAb, 2.4G2) (**b**) and CR1 and CR2 (mAb 7G6) (**d**) is shown by indirect immunoperoxidase staining. Note that Fc-mediated trapping is restricted to the light zone (*L*), whereas complement-mediated trapping also occurs in the dark zone. *M*, mantle zone; *P*, periarteriolar lymphocyte sheath. (Reproduced from YOSHIDA et al. 1993, with permission; x 200)

the absence of complement. This binding was prevented by an mAb against FcγRII. FcγRII is also expressed on FDC in draining peripheral lymph nodes of immunized animals (KOSCO et al. 1986; MAEDA et al. 1992), and isolated FDC from stimulated peripheral lymph nodes are able to bind immune complexes in the absence of complement (BRAUN et al. 1987; HEINEN et al. 1986). The latter experiments also demonstrated that murine IgG_1, IgG_{2a}, and IgG_{2b} complexes bound much better than IgM or IgG_3 complexes, and this matches nicely with the specificity of FcγRII in the mouse (SEGAL and TITUS 1978). The presence of FcγRII in mesenteric lymph nodes of unstimulated mice (MAEDA et al. 1992) does not necessarily contradict our finding that FDC in spleen from normal animals lack FcγRII, since mesenteric lymph nodes are suspected to be subject to a significant degree of antigenic stimulation under conventional conditions. The same holds true for FDC isolated from human tonsils, which also exhibit Fc-mediated binding of immune complexes (Heinen et al. 1985; Simar et al. 1984) and express FcγRII (Petrasch et al. 1990). The fact that others did not detect FcγRII on isolated FDC (Schriever et al. 1989; Sellheyer et al. 1989) might indicate that their FDC preparations only contained the FcγRII-negative subpopulation, which would explain the absence of IgG (Schriever et al. 1989) and CD23 (Sellheyer et al. 1989). The induction of FcγRII is stimulated animals may also account for the residual binding of in vivo injected immune complexes to FDC after CoVF treatment (Tew et al. 1979; Phipps et al. 1980) and for the trapping of colloidal gold-labeled immune complexes, which are unable to fix C3 (Radoux et al. 1985), because all of these studies were performed with immunized animals.

The trapping of IgE immune complexes is only poorly documented. Moderate amounts of IgE have been found in association with FDC in human tissues (Imai et al. 1986). In contrast to other classes of antibody, it seems unlikely that complement receptors play a major role in the trapping of IgE complexes, because IgE complexes do not activate complement (Farkas et al. 1982). FDC also lack the high-affinity receptor for IgE (FcεRI) found on mast cells. However, human FDC in secondary follicles do express the low-affinity receptor FcεRII (CD23), and it seems likely that this is the main receptor for IgE immune complexes on FDC. Recent findings in the mouse indicate that this is indeed the case. There are only very low levels of FcεRII on FDC from unimmunized animals, but their level is greatly enhanced by antigenic stimulation, and this correlates with increased binding of IgE. Treatment of these animals with anti-FcεRII mAb prevented the trapping of IgE immune complexes by FDC, strongly suggesting that the FcεRII does indeed mediate IgE immune complex binding to FDC.

It also emerged from these studies that the regulation of FcεRII expression on FDC was different from that on B cells, since it was also upregulated during a typical IgG response, whereas an increase in B cell FcεRII was only detected after induction of an IgE response. The relevance of this is illustrated by IgE-hypersensitive patients that maintain high levels of circulating IgE, and have high levels of IgE and FcεRII on their FDC (Mitani et al. 1988). Another interesting point is that FcεRII, like FcγRII (CDw32), appears to be restricted to the FDC located in the apical light zone of the secondary follicle (Johnson et al. 1986). Taken together, these findings indicate that during the germinal center reaction a subpopulation of FDC is induced to express FcγRII and FcεRII, which allows trapping of immune complexes containing IgG and IgE, respectively. The significance of this in the regulation of the immune response remains to be established. One possibility is that Fc receptors mediate the long-term retention of immune complexes (Mandel et al. 1981), and this would explain why carbon particles, which are probably only trapped via complement receptors, are more rapidly cleared from the follicles than antibody-containing complexes (Groeneveld et al. 1983). Obviously, this will need further investigation.

2.3.3 Development of the Immune Complex Trapping Capacity of Follicular Dendritic Cells

During the ontogeny of the lymphoid tissue, the first immune complex trapping is observed on so-called fibroblastic reticulum cells localized in primary follicles as they develop (DIJKSTRA et al. 1982, 1983), and these soon differentiate into mature FDC. It seems likely that trapping by these cells is mediated via complement receptors (see Sect. 2.3.1). Severe combined immunodeficient (SCID) mice lack the capacity to trap immune complexes in vivo (KAPASI et al. 1993; YOSHIDA et al. 1994), and this is at least in part due to the fact that these animals lack FDC that express CR1 or CR2 (YOSHIDA et al. 1994). Reconstitution with lymphocytes induces the differentiation of (host-derived) FDC that express functional CR1 and CR2 and restores the capacity to trap immune complexes (YOSHIDA et al. 1994). It seems that B cells, and not T cells, are responsible for this, as trapping can still occur in the absence of T cells (KLAUS and HUMPHREY 1977; MJAALAND and FOSSUM 1987). This idea is also in line with the absence of follicular trapping and lack of mature FDC in animals that have been depleted of B cells by neonatal anti-IgM treatment (ENRIQUEZ-RINCON et al. 1984; MACLENNAN and GRAY 1986; CERNY et al. 1988). Thus, B cells may not only play a role in the transport of immune complexes to the FDC (see Sect. 2.1), but may also be required for the differentiation of local precursor cells into mature FDC that express CR1 and CR2, which thereby acquire the capacity to trap immune complexes.

It should be mentioned that the origin of the "early" precursor of the FDC is still not clear. The original experiments by HUMPHREY and SUNDARAM (1984) using transfer of bone marrow cells into irradiated recipients demonstrated that, at least in adult animals, FDC are not bone marrow derived. We have recently confirmed this by reconstitution of adult SCID mice (YOSHIDA et al. 1994), thereby ruling out the possibility that the failure to detect possible donor-derived FDC could be due to an extremely slow turnover of mature FDC. In addition to this, IMAZEKI et al. (1992), studying the regeneration of implanted spleen, provided important evidence for the local origin of FDC. The experiments carried out by KAPASI et al. (1993), however, indicate that transfer into newborn SCID mice generates FDC of bone marrow origin. This may indicate that, early during postnatal development, a population of early precursor cells leaves the bone marrow to populate the lymphoid tissues (and/or other tissues), which will function as a local self-replenishing reservoir of fibroblast-like FDC precursors, which in turn, in the presence of B cells, will differentiate into mature FDC.

3 Role of Immune Complexes in B Cell Activation

As will be clear from the above, FDC retain antigen on their surface in the form of immune complexes, i.e., in physical association with antibody and C3 fragments.

This is important since B cells, in addition to their surface immunoglobulin (sIg), also express Fc receptors and complement receptors CR1 and CR2. Therefore, interaction of B cells with an immune complex may bring these receptors in close proximity to sIg in the plasma membrane and modulate signal transduction and the B cell response.

3.1 Role of Fc Receptors

It is well documented that specific IgG antibodies can suppress the primary antibody response and the development of B cell memory by a complement-independent mechanism (reviewed by HEYMAN 1990), and it has been suggested that this is due to a direct effect on B cells by co-cross-linking of sIg and FcR (SINCLAIR and PANOSKALTIS 1987). Essentially, all murine B cells carry FcγRII, which binds immune complexes of IgG_1, IgG_{2a}, and IgG_{2b} (DICKLER 1976). It has now become clear that co-cross-linking of sIg and FcγRII on B cells in vitro causes abortive B cell activation. Compared to ligation of sIg alone, it profoundly inhibits the proliferation induced via sIgM or sIgD, whereas the increased expression of major histocompatibility complex (MHC) class II is hardly affected (KLAUS et al. 1984). In addition, co-cross-linking FcγRII to sIg has been shown to cause an early shutdown of inositol 3-phosphate (IP_3) generation (BIJSTERBOSCH and KLAUS 1985), and this appears to be due to uncoupling of sIg from their associated G protein (RIGLEY et al. 1989). On the other hand, the intracellular Ca^{2+} release and Ca^{2+} influx remained unaffected, suggesting that the short-term release of IP_3 is sufficient to induce the Ca^{2+} response (KLAUS et al. 1987), which in turn is believed to result in the increased expression of MHC class II.

It should be noted that IgG-mediated suppression in vivo is only observed when particulate antigens are used that are able to cause sufficient cross-linking of FcR (HEYMAN 1990). Since particulate antigens are rapidly degraded and immune complexes on FDC probably contain particle-derived soluble antigens (ENRIQUEZ-RINCON and KLAUS 1984), which are known to enhance memory by a complement-dependent mechanism (HEYMAN 1990), it seems unlikely that IgG-mediated suppression occurs in the follicles.

3.2 Role of Complement Receptors

It is well established that complement factor C3 plays an important regulatory role in the immune response. In animals depleted of C3 by CoVF (PAPAMICHAEL et al. 1975; KLAUS and HUMPHREY 1977) or in C3-deficient animals (BÖTTGER et al. 1986), there is essentially no development of B memory cells, no class switching, and therefore no secondary antibody response. Similar results have been obtained by treatment of animals with soluble CR2 or the anti-CR1/2 mAb (HEBELL et al. 1991; HEYMAN et al. 1990; WIERSMA et al. 1991; THYPHRONITIS et al. 1991). Whether this is merely caused by the absence of follicular trapping of antigen or whether immune complex-associated C3 fragments (localized on FDC or elsewhere) do indeed provide a secondary signal for B memory differentiation in vivo (DUKOR and HARTMANN 1973; KLAUS and HUMPHREY 1986), or both, remains to be established. The ability of polyvalent C3 fragments to enhance the proliferation of B cells in vitro has been demonstrated (MELCHERS et al. 1985; ERDEI et al. 1985; TSOKOS et al. 1990).

Essentially all B cells express complement receptors CR1 and CR2, and it has now become clear that ligation of these receptors, in particular CR2, is able to modulate B cell activation. The role of CR2 in the activation of human B cells is well documented. In co-cross-linking experiments, anti-CR2, but not anti-CR1, has been shown to synergize with suboptimal concentrations of anti-Ig in proliferation and Ca^{2+} mobilization (CARTER et al. 1988). It has recently become clear that part of the CR2 molecules are associated with other components in the so-called CD19–CR2–TAPA-1 complex (reviewed by FEARON 1993), in which CR2 functions as the receptor for extracellular ligands, whereas CD19 transduces the signals (CARTER et al. 1991; TUVESON et al. 1993; MATSUMOTO et al. 1993; UCKUN et al. 1993) and thereby allows the B cell to respond to extremely low doses of antigen (CARTER and FEARON 1992). It should be mentioned that there are several other ligands for CR2, including EBV gp350/220 (TANNER et al. 1987), interferon-α (DELCAYRE et al. 1991), and FcεRII (CD23; AUBRY et al. 1992), but it seems likely that the C3d binding is the most relevant, because only immune complexes containing C3d can be expected to cause co-cross-linking of sIg and the CD19–CR2–TAPA-1 complex, which is essential for the induction of proliferation (CARTER and FEARON 1992) and which maintains the clonal specificity of the system.

Our recent experiments have shown that also in the mouse anti-CR1 and anti-CR1/2 synergize with suboptimal anti-IgM or IgD in the intracellular release of Ca^{2+} (VAN DEN BERG and KLAUS, submitted). This suggests that the mouse CR1 and CR2 both have signaling properties similar to human CR2, which is compatible with their close homology to human CR2 (HOLERS et al. 1992). Interesting, the signal generated by coligation of sIg and complement receptors is qualitatively different from that generated by sIg ligation alone, since it is much less sensitive to feedback regulation by protein kinase C activation (VAN DEN BERG and KLAUS, submitted), as was previously shown for human B cells (CARTER et al. 1991). Whether murine CR1 and CR2 are indeed associated with the murine homologue of CD19 remains to be determined.

4 Role of Immune Complexes on Follicular Dendritic Cells in the Memory Response

After the first administration of an antigen, a primary immune response is initiated. The first antibody-forming cells are observed in the T-dependent areas of the lymphoid tissues, and most of them secrete antibodies of the IgM class (VAN ROOIJEN et al. 1986). As soon as antibodies appear in the circulation, immune complexes will be formed, and a small proportion of these are trapped on the surface of FDC in the follicles of the lymphoid tissues. About 1 week after antigen administration, the follicles develop small focal areas, called germinal centers, that contain rapidly proliferating B lymphocytes. Evidence suggests that immune complexes are not absolutely required for the development of the germinal center (KROESE et al. 1986), although they may accelerate their development (KUNKL and KLAUS 1981). On the other hand, the generation of the germinal center is strictly T cell dependent and it is known that B cells require preactivation, probably by interaction with T cells. This may occur in the T cell areas of the lymphoid tissues, where activated T cells interact with B cells and secrete cytokines, such as interleukin-4 and interferon-γ, that are known to induce class switching (VAN DEN EERTWEGH et al. 1993). The B cells in the germinal center undergo somatic mutation in the variable region of their Ig genes, which generates a pool of cells bearing variant receptors with different affinities for antigen (JACOB et al. 1991). Whether this process of somatic hypermutation is also initiated in the T cell areas is not known. The variants with low affinity for antigen will go into apoptosis and will be ingested by the locally present tingible body macrophages, whereas high-affinity ligands will be selected and give rise to long-lived memory B Cells. These memory cells, upon secondary antigenic stimulation, will differentiate into antibody-secreting plasma cells. It is clear that the selection of high-affinity B cells requires antigen, and it seems highly likely that this occurs locally, i.e., by using the antigen that is presented by the FDC. It also seems that this antigen, which can be retained for a very long period, is used for periodical restimulation of memory B cells, as memory rapidly wanes in the absence of antigen (GRAY and SKARVAL 1988); this is important for the maintenance of the antibody response (TEW et al. 1980).

The localization of an immune complex on FDC is not enough for memory induction. Thymus-independent antigens and nonimmunogenic compounds can also be trapped by FDC, but this does not result in memory formation (DRESSER and POPHAM 1979). This is because interactions with T-helper cells are required. Whether these only occur outside the follicles or whether interactions with T cells in the germinal center are also required is not known. Interestingly, activated T cells in both the T cell area and in the follicles express the ligand for CD40 (VAN DEN EERTWEGH et al. 1993), and ligation of CD40 in combination with sIg triggering has been shown to rescue germinal center B cells from apoptosis in vitro (LIU et al. 1989); this may be relevant for the selection of high-affinity B cell clones in vivo.

The generation of memory is C3 dependent, but the exact role of C3 can only be explained in part. As will be obvious from the above, C3 is required for the trapping of immune complexes on the FDC, which is crucial for the selection of high-affinity B cell clones. Apart from this, C3 fragments associated with antigen may play a direct role in B cell activation. The fact that C3 depletion prevents the generation of germinal centers (WHITE et al. 1975) and that this is not simply due to the absence of antigen (KROESE et al. 1986, 1990) suggests that C3 in combination with antigen may function as a proliferation signal for germinal center B cells by co-cross-linking sIg and the CR2–CD19–TAPA-1 complex and that this contributes to the efficiency of the affinity maturation process. This may be of particular importance at suboptimal doses of antigen, because then the C3 dependence and the effect of anti-complement receptor antibodies is most pronounced (HEYMAN et al. 1988,1990; WIERSMA et al. 1991).

5 Conclusions

The follicular dendritic cell is a major constituent of the microenvironment of the lymphoid follicle. It has the characteristic property of trapping antigen in the form of immune complexes composed of antigen–antibody–C3 and to retain these complexes on its surface. CR1 and CR2 mediate the binding of immune complexes to the FDC in normal animals, whereas in stimulated animals the binding of IgG and IgE can also occur via FcγRII and FcεRII respectively. The antigen trapped on the surface of the FDC plays a crucial role in the generation of B cell memory, by selecting high-affinity mutants from B cells undergoing somatic mutation in the germinal center. C3 fragments associated with immune complexes not only play a crucial role in trapping immune complexes by FDC, but can also act as an important costimulus for B cell proliferation in the germinal center by co-cross-linking surface Ig and complement receptors.

References

Aubry JP, Pochon S, Graber P, Jansen KU, Bonnefoy JY (1992) CD21 is a ligand for CD23 and regulates IgE production. Nature 358: 505–507

Böttger EC, Metzger S, Bitter-Suermann D, Stevenson G, Kleindienst S, Burger R (1986) Impaired humoral immune response in complement C3-deficient guinea pigs: absence of secondary antibody response. Eur J Immunol 16: 1231–1235

Braun M, Heinen E, Cormann N, Kinet-Donoël C, Simar LJ (1987) Influence of immunoglobulin isotypes and lymphoid cell phenotype on the transfer of immune complexes to follicular dendritic cells. Cell Immunol 107: 99–106

Brown JC, DeJesus DG, Holborow EJ, Harris G (1970) Lymphocyte mediated transport of aggregated human γ-globulin into germinal centre areas of normal mouse spleen. Nature 228: 367–369

Bijsterbosch MK, Klaus GGB (1985) Crosslinking of surface Ig and Fc receptors on B lymphocytes inhibits stimulation of inositol phospholipid breakdown via the antigen receptors. J Exp Med 162: 1825

Carter RH, Fearon DT (1992) CD19: lowering the threshold for antigen receptor stimulation of B lymphocytes. Science 256: 105–107

Carter RG, Spycher MO, Yin CNG, Hoffman R, Fearon DT (1988) Synergistic interaction between complement receptor type 2 and membrane IgM on B lymphocytes. J Immunol 141: 457–463

Carter RG, Tuveson DA, Park DJ, Rhee SG, Fearon DT (1991) The CD19 complex of lymphocytes: activation of phospholipase C by a protein tyrosine kinase-dependent pathway that can be enhanced by the membrane IgM complex. J Immunol 147: 3663–3671

Cerny A, Zinkernagel RM, Groscurth P (1988) Development of dendritic cells in lymph nodes of B-cell-depleted mice. Cell Tissue Res 254: 449–454

Chen LL, Frank AM, Adams JC, Steinman RM (1978) Distribution of horseradish peroxidase (HRP)-anti-HRP immune complexes in mouse spleen with special reference to follicular dendritic cells. J Cell Biol 79: 184–199

Collie MH (1974) The localization of soluble antigen in the spleen of Xenopus laevis. Experientia 30: 1205–1206
Damoiseaux JGMC, Döpp EA, Neefjes, Beelan RHJ, Dijkstra CD (1989) Heterogeneity of macrophages in the rat evidenced by variability in determinants. Two new anti-rat macrophage antibodies against a heterodimer of 160-kD and 95-kD (CD11b/CD18). J Leukocyte Biol 46: 556 –564
Delcayre AX, Salas F, Mathur S, Kovats K, Lotz M, Lernhardt W (1991) Epstein-Barr virus/complement C3d receptor is an interferon α receptor. EMBO J 10: 919–926
Dickler HB (1976) Lymphocyte receptors for immunoglobulin. Adv Immunol 24: 167
Dijkstra CD, Van Tilburg NJ, Döpp EA (1982) Ontogenetic aspects of immune-complex trapping in the spleen and popliteal lymph nodes of the rat. Cell Tissue Res 223: 545–552
Dijkstra CD, Te Velde AA, Van Rooijen N (1983) Localization of horseradish (HRP)-anti-HRP complexes in cryostat sections: influence of endotoxin on trapping of immune complexes in the spleen of the rat. Cell Tissue Res 232: 1–7
Dresser DW, Popham AM (1979) The influence of T cells on the initiation and expression of immunological memory. Immunology 38: 265
Dukor P, Hartmann KU (1973) Hypothesis: bound C3 as a second signal for B cell activation. Cell Immunol 7: 349–356
Eikelenboom P, Kroese FGM, Van Rooijen N (1983) Immune complex-trapping cells in the spleen of the chicken. Cell Tissue Res 231: 377–386
Embling PH, Evans H, Guttierez C, Holborow EJ, Johns P, Johnson PM, Papamichail M, Stanworth DR (1978) Structural requirements for immunoglobulin aggregates to localize in germinal centres. Immunology 34: 781–786
Enriquez-Rincon F, Klaus GGB (1984) Follicular trapping of hapten-erythrocyte-antibody complexes in mouse spleen. Immunology 52: 107–116
Enriquez-Ricon F, Andrew E, Parkhouse RME, Klaus GGB (1984) Suppression of follicular trapping of antigen-antibody complexes in mice treated with anti-IgM or anti-IgD antibodies from birth. Immunology 53: 713–719
Erdei A, Melchers F, Schultz T, Dierich M (1985) The action of human C3 in soluble or cross-linked form with resting and activated murine B lymphocytes. Eur J Immunol 15: 184–188
Farkas AI, Medgyesi GA, Füst G, Miklos K, Gergely J (1982) Immunogenicity of antigen complexed with antibody. I. Role of different isotypes. Immunology 45: 483–492
Fearon DT (1993) The CD19–CR2–TAPA-1 complex, CD45 and signalling by the antigen receptor of B lymphocytes. Curr Opinion Immunol 5: 341–348
Fearon DT, Ahearn JM (1989) Complement receptor type 1 (C3b/C4b receptor; CD35) and complement receptor type 2 (C3d/Epstein-Barr Virus receptor; CD21). In: Lambris JD (ed) The third component of complement. Chemistry and biology. Springer, Berlin Heidelberg New York, pp 83–98 (Current topics in microbiology and immunology, vol 153)
Fingeroth JD (1990) Comparative structure and evolution of murine CR2. J Immunol 144: 3458–3467
Fingeroth JD, Benedict MA, Levy DN, Strominger JL (1989) Identification of murine complement receptor type 2. Proc Natl Acad Sci USA 86: 242–246
Gajl-Peczalaska KJ, Fish AJ, Meuwisse HJ, Frommel D, Good RA (1969) Localization of immunological complexes fixing beta1c (C3) in germinal centers of lymph nodes. J Exp Med 130: 1367–1374
Gerdes J, Stein H, Mason DY, Ziegler A (1983) Human dendritic reticulum cells of lymphoid follicles: their antigenic profile and their identification as multinucleated giant cells. Virchows Arch B 42: 161–172
Gray D, Skarvall H (1988) B cell memory is short-lived in the absence of antigen. Nature 336: 70–73
Gray D, McConnel I, Kumararatne DS, MacLennan ICM, Humphrey JH, Bazin H (1984a) Marginal zone cells express CR1 and CR2 receptors. Eur J Immunol 14: 40–52
Gray D, Kumararatne DS, Lortan J, Khan M, MacLennan ICM (1984b) Relation of intrasplenic migration of marginal zone B cells to antigen localization of follicular dendritic cells. Immunology 52: 659–669
Groeneveld PHP, Eikelenboom P, Van Rooijen N (1983) Mechanism of follicular trapping: similarities and differences in trapping of antibody-complexed antigens and carbon particles in the follicles of the spleen. J Reticuloend Soc 33: 109–117
Halstensen TS, Mollnes TE, Brandtzaeg P (1988) Terminal complement complex (TCC) and S-protein (vitronectin) on follicular dendritic cells in human lymphoid tissues. Immunology 65: 193–197
Hebell T, Ahearn JM, Fearon DT (1991) Suppression of the immune response by a soluble complement receptor. Science 254: 102–105
Heinen E, Radoux D, Kinet-Denoël C, Moeremans M, De Mey J, Simar LJ (1985) Isolation of follicular dendritic cells from human tonsils and adenoids. III. Analysis of their Fc receptors. Immunology 54: 777–784

Heinen E, Braun M, Coulie PG, Van Snick J, Moeremans M, Cormann N, Kinet-Denoël C, Simar LJ (1986a) Transfer of immune complexes from lymphocytes to follicular dendritic cells. Eur J Immunol 16: 167–172

Heinen E, Cormann N, Kinet-denoël C, Simar L (1986b) Lipopolysaccharide suppresses immune complex retention by FDC without cytological alterations. Immunol Lett 13: 323–327

Heyman B (1990) The immune complex: possible ways of regulating the antibody response. Immunol Today 11: 310–313

Heyman B, Pilstrom L, Shulman MJ (1988) Complement activation is required for IgM-mediated enhancement of the antibody response. J Exp Med 167: 1999

Heyman B, Wiersma EJ, Kinoshita T (1990) In vivo inhibition of the antibody response by a complement receptor specific monoclonal antibody. J Exp Med 172: 665–668

Holers VM, Kinoshita T, Molina H (1992) The evolution of mouse and human complement C3-binding proteins: divergence of form but conservation of function. Immunol Today 13: 231–236

Humphrey JH, Franck MM (1967) Localization of non-microbial antigens in draining lymph nodes of tolerant normal and primed rabbits. Immunology 13: 87–95

Humphrey JH, Sundaram V (1984) The origin of follicular dendritic cells in the mouse and the mechanism of trapping of immune complexes. Eur J Immunol 14: 859–864

Imai Y, Yamakawa M, Masuda A, Sato T, Kasajima T (1986) Function of the follicular dendritic cell in the germinal center of lymphoid follicles. Histol Histopathol 1: 341–353

Imazeki N, Senoo A, Fuse Y (1992) Is the follicular dendritic cell a primarily stationary cell? Immunology 76: 508–510

Jacob J, Kelsoe G, Rajewski K, Weiss U (1991) Intraclonal generation of antibody mutants in germinal centers. Nature 354: 389–392

Johnson GD, Hardie DL, Ling NR, MacLennan ICM (1986) Human follicular dendritic cells (FDC): a study with monoclonal antibodies (MoAb). Clin Exp Immunol 64: 205–213

Kamperdijk EWA, Dijkstra CD, Döpp EA (1987) Transport of immune complexes from the subcapsular sinus into the lymph node follicle of the rat. Immunology 147: 395

Kapasi ZF, Burton GF, Schultz LD, Tew TG, Szakal AK (1993) Induction of functional follicular dendritic cells development in severe combined immunodeficiency mice. Influence of B and T cells. J Immunol 150: 2648

Kapasi ZE, Kosco MH, Schultz LD, Tew JG, Szakal AK (1994) Cellular origin of follicular dendritic cells. Proceedings of the 11th germinal centre conference (in press)

Kinoshita T, Takeda J, Hong K, Kozono H, Sakai H, Inoue K (1988) Monoclonal antibodies to mouse complement receptor type 1. J Immunol 140: 3066–3072

Kinoshita T, Thyphronitis G, Tsokos GC, Finkelman FD, Hong K, Sakai H, Inoue K (1990) Characterization of murine complement receptor type 2 and its immunological cross-reactivity with type 1 receptor. Int Immunol 2: 651–659

Klaus GGB (1979) The generation of memory cells. III. Antibody class requirements for the generation of B memory cells. Immunology 37: 345–351

Klaus GGB, Humphrey JH (1977) The generation of memory cells. I. The role of C3 in the generation of memory cells. Immunology 33: 31–40

Klaus GGB, Humphrey JH (1986) A re-evaluation of the role of C3 in B cell activation. Immunol Today 7: 163–165

Klaus GGB, Pepys MB, Kitajima K, Askonas BA (1979) Activation of mouse complement by different classes of mouse antibody. Immunology 38: 687

Klaus GGB, Humphery JH, Kunkl A, Dongworth DW (1980) The follicular dendritic cell: its role in antigen presentation in the generation of immunological memory. Immunol Rev 53: 3–28

Klaus GGB, Hawrylowicz CM, Holman M, Keeler KD (1984) Intact (IgG) anti-Ig antibodies activate B cells but inhibit the induction of DNA synthesis. Immunology 53: 693

Klaus GGB, Bijsterbosch MK, O'Garra A, Harnett MM, Rigley KP (1987) Receptor signalling and crosstalk in B lymphocytes. Immunol Rev 99: 19–83

Kosco MH, Tew JG, Szakal AK (1986) Antigenic phenotyping of isolated and in situ rodent follicular dendritic cells (FDC) with emphasis on the ultrastructural demonstration of Ia antigens. Anat Rec 215: 201–213

Kroese FGM, Van Rooijen N (1983) Antigen trapping in the spleen of the turtle, Chrysemys scripta elegans. Immunology 49: 61–68

Kroese FGM, Leceta J, Döpp EA, Herraez MP, Nieuwenhuis P, Zapata A (1985) Dendritic immune complex cells in the spleen of the snake, Python reticularis. Dev Comp Immunol 9: 641–652

Kroese FGM, Wubbema AS, Nieuwenhuis P (1986) Germinal centre formation and follicular trapping in the spleen of lethally X-irradiated and reconstituted rats. Immunology 57: 99–104

Kroese FGM, Timens W, Nieuwenhuis P (1990) Germinal center reaction and B lymphocytes: morphology and function. Curr Top Pathol 84(1): 103–148
Kunkl A, Klaus GGB (1981) The generation of memory cells. IV. Immunization with antigen-antibody complexes accelerates the development of memory cells, the formation of germinal centres and the maturation of antibody in the secondary response. Immunology 43: 371–378
Kurtz CB, Paul MS, Aegerter M, Weis JJ, Weis JH (1989) Murine complement receptor gene family. II. Identification and characterization of the murine homolog (Cr2) to human CR2 and its molecular linkage to Crry. J Immunol 143: 2058–2067
Laman JD, Kors N, Van Rooijen N, Claassen E (1990) Mechanism of follicular trapping: localization of immune complexes and cell remnants after elimination and repopulation of different spleen cell populations. Immunology 71: 57–62
MacLennan ICM, Gray D (1986) Antigen-driven selection of virgin and memory B cells. Immunol Rev 91: 61–85
Maeda K, Burton GF, Padgett DA, Conrad DH, Huff TF, Masuda A, Szakal AK, Tew JG (1992) Murine follicular dendritic cells and low affinity Fc receptors for IgE (FcεRII). J Immunol 148: 2340–2347
Mandel TE, Phipps RP, Abbot AP, Tew JG (1981) Long-term antigen retention by dendritic cells in the popliteal lymph node of immunized mice. Immunology 43: 353–361
Matsumoto AK, Martin DR, Carter RH, Klickstein LB, Ahearn JM, Fearon DT (1993) Functional dissection of the CD21/CD19/TAPA-1/Leu-13 complex of B lymphocytes. J Exp Med 178: 1407–1417
Melchers F, Erdei A, Schultz T, Dierich MP (1985) Groth control of activated, synchronized murine B cells by the C3d fragment of human complement. Nature 317: 264–265
Mitani S, Takagi K, Oka T, Mori S (1988) Increased immunoglobulin E Fc receptor bearing cells in the germinal centers of hyperimmunoglobulinemia E patients. Int Arch Allergy Immunol 87: 63
Mjaaland S, Fossum S (1987) The localization of antigen in lymph node follicles on congenitally athymic nude rats. Scand J Immunol 26: 141–147
Molina H, Wong W, Kinoshita T, Brebber C, Foley S, Holers VM (1992) Distinct receptor and regulatory properties of recombinant mouse complement receptor 1 (CR1) and Crry, the two genetic homologues of human CR1. J Exp Med 175: 121–129
Nossal GJV, Abbot A, Mitchell J, Lummus Z (1968) Antigens in immunity XV. Ultrastructural features of antigen capture in primary and secondary lymphoid follicles. J Exp Med 127: 227–235
Papamichail M, Gutierrez C, Embling P, Johnson P, Holborow EJ, Pepys MB (1975) Complement dependence of localization of aggregated IgG in germinal centres. Scand J Immunol 4: 343–347
Petrasch S, Perez-Alvarez C, Schmitz J, Kosco M, Brittinger G (1990) Antigenic phenotyping of human follicular dendritic cells isolated from nonmalignant and malignant lymphatic tissue. Eur J Immunol 20: 1013–1018
Phipps RP, Mitchell GF, Mandel TE, Tew JG (1980) Antibody isotypes mediating antigen retention in passively immunized mice. Immunology 40: 459–466
Phipps RP, Mandel TE, Tew JG (1981) Effect of immunosuppressive agents on antigen retained in lymphoid follicles and collageneous tissue of immune mice. Cell Immunol 57: 505–516
Pramoonjago P, Takeda J, Kim YU, Inoue K, Kinoshita T (1993) Ligand specificities of mouse complement receptor types 1 (CR1) and 2 (CR2) purified from spleen cells. Int Immunol 5: 337–343
Radoux D, Kinet-Denoël C, Heinen E, Moeremans M, De Mey J, Simar LJ (1985) Retention of immune complexes by Fc receptors on mouse follicular dendritic cells. Scand J Immunol 21: 345–353
Reynes M, Aubers JP, Cohen JHM, Audouin J, Tricottet V, Diebold J, Kazatchkine MD (1985) Human follicular dendritic cells express CR1, CR2, and CR3 complement receptor antigens. J Immunol 135: 2687–2694
Rosen H, Law SKA (1989) The leukocyte surface receptor(s) for the iC3b product of complement. In: Lambris JD (ed) The third component of complement. Chemistry and biology. Springer, Berlin Heidelberg New York, pp 99–122 (Current topics in microbiology and immunology, vol 153)
Schriever F, Freedman AS, Freeman, Messner E, Lee G, Daley J, Nadler LM (1989) Isolated human follicular dendritic cells display a unique antigenic phenotype. J Exp Med 169: 2043–2058
Secombes CJ, Manning MJ, Ellis AE (1982) Localization of immune complexes and heat-aggregated immunoglobulin in the carp Cyprinus carpio L. Immunology 47: 101–105
Segal DM, Titus JA (1978) The subclass specificity for the binding of murine myeloma proteins to macrophage and lymphocyte cell lines and to normal spleen cells. J Immunol 120: 1395
Sellheyer K, Schwarting R, Stein H (1989) Isolation and antigenic profile of follicular dendritic cells. Clin Exp Immunol 78: 431–436
Simar LJ, Lilet-Leclercq C, Heinen E, Kinet-Denoël C, Radoux D (1984) Functional study of human tonsillar follicular dendritic cells. Acta Otorhinolaryngol Belg 38: 278–287

Sinclair NRS, Panoskaltis A (1987) Immunoregulation by Fc signals. A mechanism for self-nonself discrimination. Immunol Today 8: 76
Szakal AK, Holmes KL, Tew JG (1983) Transport of immune complexes from the subcapsular sinus to lymph node follicles on the surface of non-phagocytotic cells, including cells with dendritic morphology. J Immunol 131: 1714
Szakal AK, Haley ST, Caffrey RE, Burton GF, Tew JG (1993) Lymph node antigen transport cell (ATC) phenotype. Abstract book of the 11th Germinal Centre Conference, 1993, Spa, Belgium, p 122
Tanner J, Weis J, Fearon D, Whang Y, Kieff E (1987) Epstain-Barr virus gp350/220 binding to the B lymphocyte C3d receptor mediates absorption, capping, and endocytosis. Cell 50: 203–213
Tew JG, Mandel TE, Miller GA (1979) Immune retention: immunological requirements for maintaining an easily degradable antigen in vivo. Aust J Exp Biol Med Sci 57: 401–414
Tew JG, Phipps RP, Mandel TE (1980) The maintenance and regulation of the humoral immune response: persisting antigen and the role of follicular antigen-binding dendritic cells as accessory cells. Immunol Rev 53: 175–201
Thyphronitis G, Kinoshita T, Inoue K, Sweinle JE, Tsokos GC, Metcalf ES, Finkelman FD, Balow JE (1991) Modulation of mouse complement receptors 1 and 2 suppresses antibody responses in vivo. J Immunol 147: 224–230
Tsokos GC, Lambris JD, Finkelman FD, Anastassiou ED, June CH (1990) Monovalent ligands of complement receptor 2 inhibit whereas polyvalent ligands enhance anti-Ig-induced human B cell intracytoplasmic free calcium concentration. J Immunol 144: 1640–1645
Tuveson DA, Ahearn JM, Matsumoto AK, Fearon DT (1991) Molecular interactions of complement receptors on B lymphocytes: a CR1/CR2 complex distinct from the CR2/CD19 complex. J Exp Med 173: 1083–1089
Tuveson DA, Carter RH, Soltoff SP, Fearon DT (1993) CD19 of B cells as a surrogate kinase insert region to bind phosphatidyl inositol 3-kinase. Science 260: 986–988
Uckun FM, Burkhardt AL, Jarvis L, Jun X, Stealey B, Dibirdik I, Meyers DE, Tuel-Ahlgren L, Bolen JB (1993) Signal transduction through the CD19 receptor during discrete developmental stages of human B cell ontogeny. J Biol Chem 268: 21172–21184
Van den Berg TK, Döpp EA, Breve JJP, Kraal G, Dijkstra CD (1989) The heterogeneity of the reticulum of rat peripheral lymphod organs identified by monoclonal antibodies. Eur J Immunol 19: 1747–1756
Van den Berg TK, Döpp EA, Daha MR, Kraal G, Dijkstra CD (1992) Selective inhibition of immune complex trapping by follicular dendritic cells with monoclonal antibodies against rat C3. Eur J Immunol 22: 957–962
Van den Eertwegh AJM, Noelle RJ, Roy M, Shepherd DM, Aruffo A, Ledbetter JA, Boersma WJA, Claassen E (1993) In vivo CD40-gp39 interactions are essential for thymus-dependent humoral immunity. I. In vivo expression of CD40 ligand, cytokines ant antibody production delineates sites of cognate T–B cell interactions. J Exp Med 178: 1555–1565
Van Rooijen N (1975) Immune complexes in the spleen: the difference between competitive inhibition of immune complex trapping in spleen follicles and inhibition by paratyphoid vaccine. Immunol 28: 1155–1162
Van Rooijen N, Claassen E, Eikelenboom P (1986) Is there a single differentiation pathway for all antibody-forming cells in the spleen. Immunol Today 7: 193–196
White RG, Henderson DC, Eslami MB, Nielsen KH (1975) Localization of a protein antigen in the chicken spleen: effects of various manipulative procedures on the morphogenesis of the germinal center. Immunology 28: 1–21
Wiersma EJ, Kinoshita T, Heyman B (1991) Inhibition of immunological memory and T-independent humoral responses by monoclonal antibodies specific for murine complement receptors. Eur J Immunol 21: 2501–2506
Yoshida K, Van den Berg TK, Dijkstra CD (1993) Two functionally different follicular dendritic cells in secondary lymphoid follicles of mouse spleen, as revealed by CR1/2 and FcγRII-mediated immune complex trapping. Immunology 80: 34–39
Yoshida K, Van den Berg TK, Dijkstra CD (1994) The functional state of follicular dendritic cells in severe combined immunodeficient (SCID) mice: role of lymphocytes. Eur J Immunol 24: 464–468

Follicular Dendritic Cells: Antigen Retention, B Cell Activation, and Cytokine Production

M.H. Kosco-Vilbois and D. Scheidegger

1 Introduction

Follicular dendritic cells (FDC) are located in dynamic microenvironments that are essential for generating effective and complete humoral responses. These anatomical locations present features of either a primary follicular nodule or a secondary structure known as the germinal center. During primary and secondary exposure to antigen, germinal centers develop within primary follicles according to a set time course (Fig. 1; Coico et al. 1983; Tew et al. 1992). The outcome of the complicated cellular interactions occurring in germinal centers is an increased frequency of antigen-experienced (i.e., memory) circulating B cells and pre-plasma cells that will finish their differentiation process elsewhere. The humoral response also matures during the germinal center reaction through the mechanisms of class switching, somatic mutation, and clonal selection (reviewed in Kosco and Gray 1992; MacLennan 1994).

The cellular and molecular components that give rise to each phase of the germinal center reaction appear to depend on cell surface receptor–ligand interactions as well as cytokine production and release. The role of FDC in these reactions is multifaceted, with new aspects continually being revealed. This chapter will attempt to describe those components that we have found to be attributable to FDC as well as to speculate on possibilities not yet confirmed.

Basel Institute for Immunology, Grenzacherstrasse 487, 4005 Basel, Switzerland
Present address: GLAXO Institute for Molecular Biology,14, Chemin des Aulx, Case Postale 674, 1228 Plan-les-Ouates, Switzerland

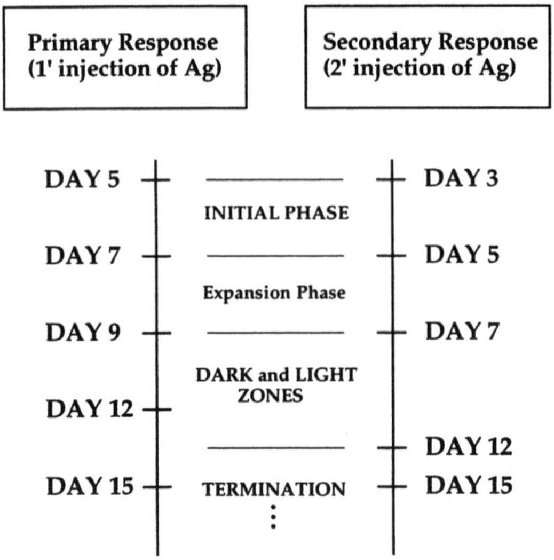

Fig. 1. Kinetics of a germinal center response after primary immunization with a protein antigen (*Ag*) precipitated in alum and following the challenge immunization with the protein in soluble form

2 Follicular Dendritic Cells as Long-Term Depots of Antigen and Pivotal Players in Maintaining Memory Responses

During the 1970s, several laboratories investigated the possibility of eliciting spontaneous antibody responses after adoptive transfer of cells from immune animals to naive hosts. This led to the observation that cells from lymphoid organs that received antigen (i.e., via drainage of the immunization site) were much more effective in producing high titers of antigen-specific immunoglobulin (Ig)G and IgE than the nondraining tissue. Since it had been shown that antigen-specific lymphocytes circulated throughout the lymphoid system, Tew and colleagues proposed that the antigen retained on FDC played a major role in the maintenance and regulation of the humoral immune response (Tew et al. 1980). Data and possible models for this concept have been extensively reviewed in two publications and will not be reexamined here (Mandel et al. 1980; Tew et al. 1980).

However, one experiment conducted by Tew and colleagues is particularly relevant to this review. This involved the surgical removal of draining versus nondraining lymphoid organs from a mouse in order to assess maintenance of specific antibody titers within the animal. Mice were primed and challenged with human serum albumin (HSA) in complete Freund's adjuvant in the front (group A) or hind (group B) limbs. Two months later, the inguinal and popliteal lymph nodes plus the spleen were excised from both groups. At 6-week intervals after surgery, anti-HSA serum titers were assessed. Their data demonstrated that removal of the tissues involved in antigen drainage and retention (group B) resulted in a 50%

decrease in specific antibody serum titers as compared to the mice in which only the nondraining tissues were excised (group A). While this relationship of unequal titers between the two groups was maintained throughout the length of the experiment (18 weeks), at no time did the specific anti-HSA level totally disappear from group B. This persistence of a lower but significant antibody titer may have been due to production by long-lived plasma cells in the bone marrow (Ho et al. 1986) and/or the possibility that antigen drained to additional sites not removed during surgery (e.g., the para-aortic lymph nodes, see below).

In addition to the influence on antibody titers, GRAY and SKARVALL (1988) published a study which linked the retention of antigen on FDC to the maintenance of memory B cell clones. This work, which has also been discussed at length in a review (GRAY and LEANDERSON 1990), demonstrated that when immune B cell populations were transferred to naive hosts, a significant decrease in the ability to mount a memory antibody response occurred within 6 weeks when no antigen was present. From this data, they proposed that, following an initial exposure, memory B cells need a periodic reencounter with antigen in order to persist. Contact with antigen would provide a survival signal that may or may not also require a second signal from a T cell (GRAY and LEANDERSON 1990). Whether entry into cell cycle and subsequent proliferation is also required for survival is still to be determined (GRAY 1992).

A study conducted recently in our laboratory was designed to address the following questions related to these observations:

1. How long can antigen be detected within the draining lymphoid tissue and which lymph nodes are involved in the antigen persistence?
2. If the first lymph node of the draining chain is surgically removed, will the decrease in antigen-specific IgG titers be significant?
3. Will memory B cell clones be maintained in the absence of this draining lymph node?

In these experiments, mice received a single subcutaneous injection of tetanus toxoid (TT) precipitated in alum in the left hind limb only; the anti-TT levels in the serum were measured at regular intervals (M.H. KOSCO-VILBOIS, D. SCHEIDEGGER, and I. ESPARZA, unpublished data). We chose to use a single injection given as an alum precipitate so that the amount of antigen would minimally diffuse along the draining lymphoid chain, yet be sufficient to provoke a reasonable antibody response.

As can be seen in Fig. 2 A–D, at 5 weeks postimmunization, immune complexes containing TT could be detected extensively along the FDC network within the draining popliteal lymph nodes. In addition, antigen was detected in the para-aortic lymph nodes, albeit less abundantly (Fig. 2F). No antigen was found further along the left side of the mouse nor in any of the lymph nodes from the nondraining side (Fig. 2E). The draining and nondraining sets of popliteal, para-aortic, inguinal, brachial, axillary, cervical, and submandibular lymph nodes from individual mice were evaluated.

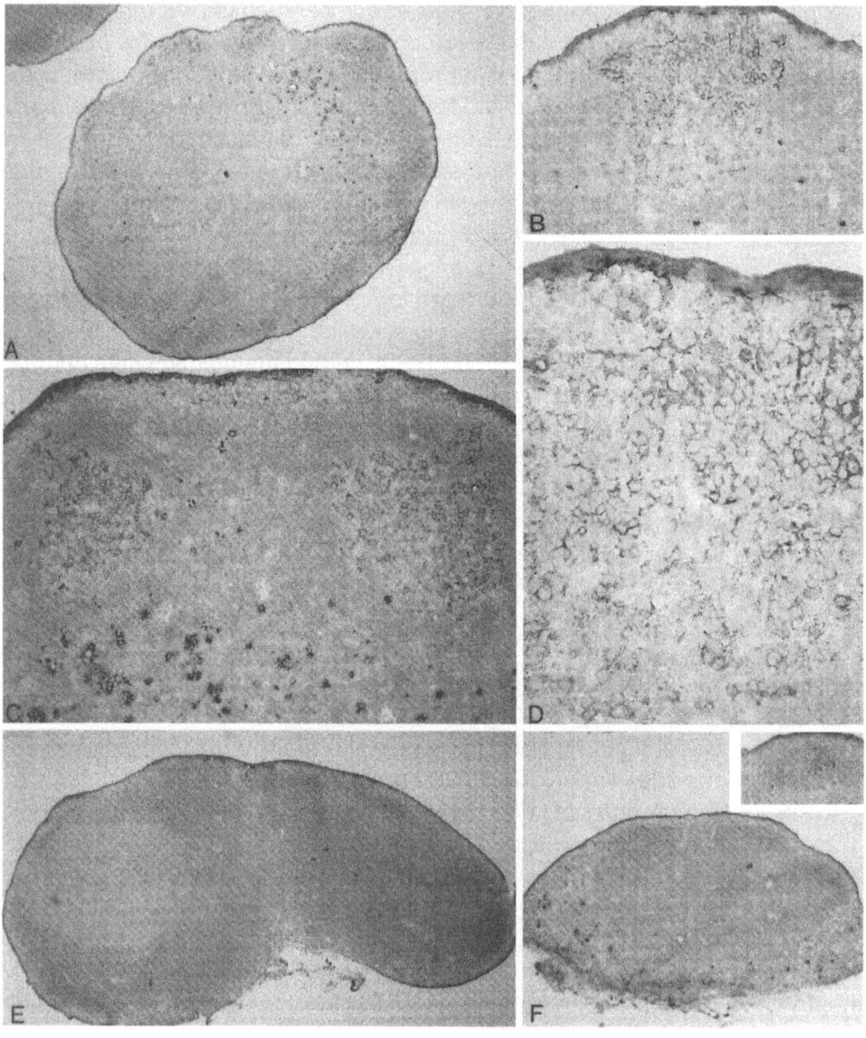

Fig. 2.A–F. Localization of tetanus toxoid in the immune complex-retaining reticulum of follicular dendritic cells (FDC) 5 weeks after primary immunization. Mice were immunized subcutaneously with 50 μg alum-precipitated tetanus in the left hind limb only. Five weeks later, all lymph nodes were collected and cryostat sections were processed for localization of antitetanus antibodies. **A–D** Numerous germinal centers containing antitetanus immune complexes were detected in the draining popliteal lymph node: low-level magnification of entire popliteal lymph node (**A**),higher magnifications of individual germinal centers (**B,C**), and higher magnification of Fig. 2B detailing the FDC network within the light and dark zones of the germinal center (**D**). Plasma cells producing antitetanus antibodies were also visualized in this lymph node (very dark labeling cells outside of germinal centers; **A,C**). **E** In most other lymph nodes from the draining or nondraining side of the lymphatic chain, no retention of tetanus or tetanus-specific plasma cells was detected; left axillary lymph node shown here. **F** The exceptions were the three para-aortic lymph nodes, where one or two also had multiple sites of antigen retention as well as tetanus-specific plasma cells; para-aortic lymph node with three individual sites of tetanus localization and darkly labeling plasma cells shown here. *Inset*, higher magnification of one of the sites of tetanus retention

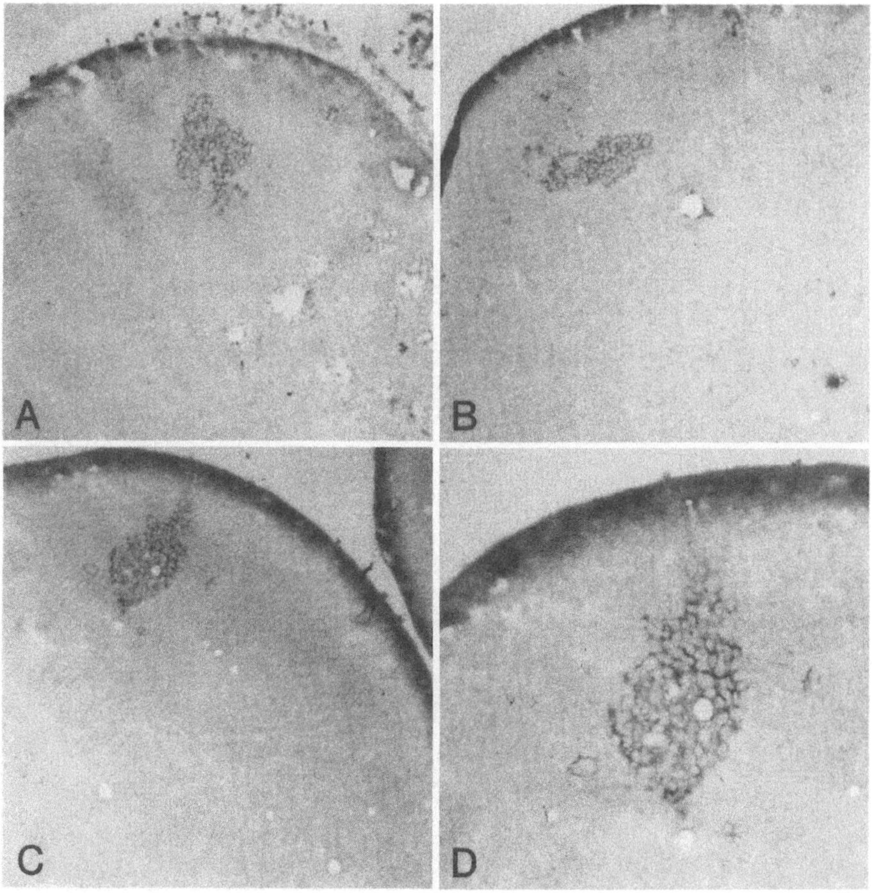

Fig. 3 A–D. Localization of tetanus toxoid in the immune complex-retaining reticulum of follicular dendritic cells (FDC) 5 months after primary immunization. Mice were immunized subcutaneously with 50 µg alum-precipitated tetanus in the left hind limb only. Five months later, all lymph nodes were collected and cryostat sections were processed for localization of antitetanus antibodies. **A–C**, Various, peculiar morphological arrangements of the immune complex retention in three different draining popliteal lymph nodes. **D** Higher magnification of Fig. 3C detailing the architecture of the immune complex retaining network

Figure 3 provides examples of the more tightly associated retention of complexes containing TT observed on the processes of FDC 5 months after immunization. In all ten mice evaluated, one to three sites of TT retention were observed per draining popliteal lymph node. In addition, at least one para-aortic lymph node contained a TT-positive FDC network in each mouse.

The longest interval after a primary immunization at which we could confidently detect the retention of TT (using light microscopy) was 14 months. Figure 4 demonstrates the visualization of either antitetanus antibodies (Fig. 4A,B) or the antigen itself (Fig. 4C, D) in a left popliteal lymph node. Using the

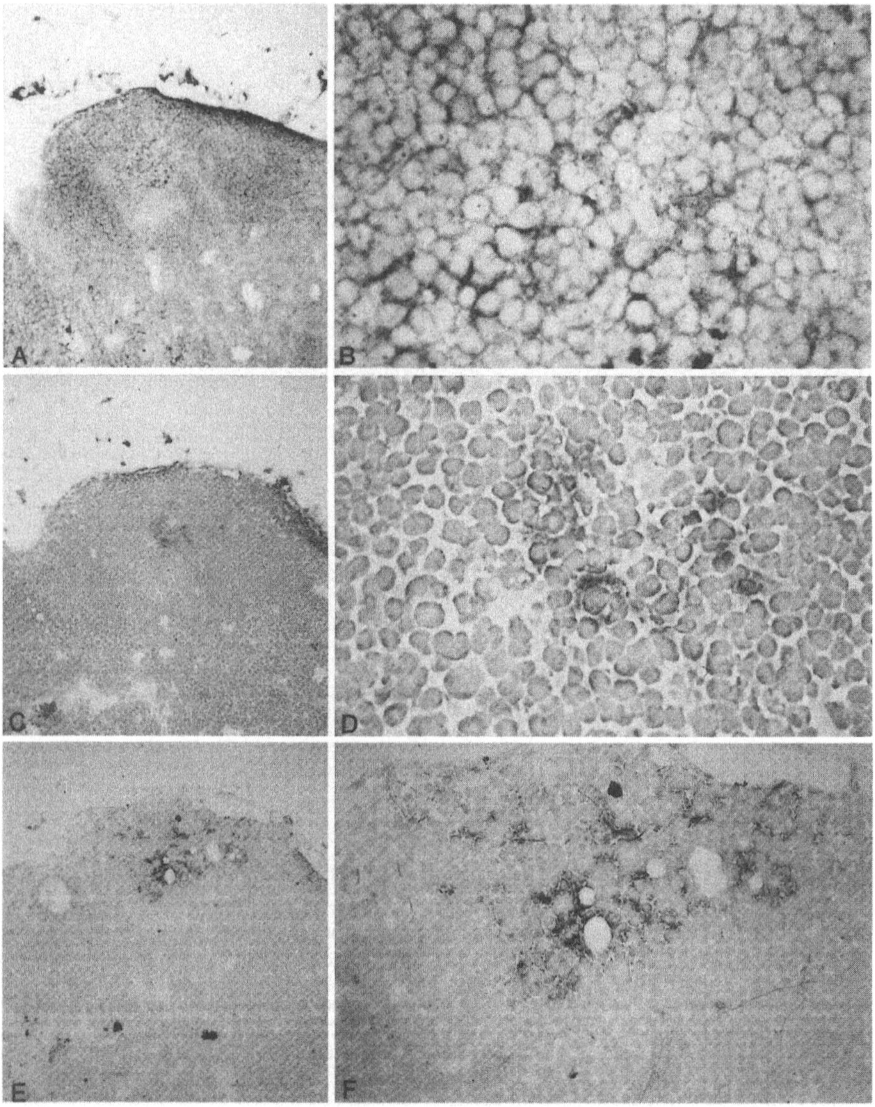

Fig. 4 A–F. Localization of tetanus toxoid in the immune complex-retaining reticulum of follicular dendritic cells (FDC) 14 months after primary immunization. Mice were immunized subcutaneously with 50 μg alum-precipitated tetanus in the left hind limb only. Fourteen months later, all lymph nodes were collected and serial cryostat sections were processed for localization of either the antitetanus antibodies, the tetanus itself, or the FDC themselves. **A** Localization of antitetanus antibodies within the immune complex-retaining reticulum (*dark area* in the center of the photomicrograph) in the light zone of a germinal center. Also note the level of antibody that appears to be "bathing" the entire tissue (i.e., light level of labeling in between cells as compared to Fig. 4C, 4E). **B** Higher magnification of Fig. 4A. **C** Localization of the antigen (i.e., tetanus) in the immune complexes in the adjacent serial section. **D** Higher magnification of Fig. 4C. **E** Confirmation that the localization of the antibody and antigen in the tetanus-specific immune complexes corresponds to the site where FDC reside. FDC were visualized using the rat anti-mouse FDC monoclonal antibody, FDC-M1. **F** Higher magnification of Fig. 4E, but an intermediate magnification between Fig. 4A/C and 4B/D

rat anti-mouse FDC-specific monoclonal antibody, FDC-M1, the location of the FDC network was visualized (Fig. 4E, F), confirming that the tetanus–antitetanus immune complexes were indeed retained by FDC.

Next, we attempted to address the relationship between retention of antigen on FDC and maintenance of serum antitetanus titers. For this, 5 months after a primary injection of TT into the left hind limb (as described above), the draining popliteal lymph node was surgically removed. Total TT antibody titers were determined after various intervals by an enzyme-linked immunosorbent assay (ELISA); the dilution at which the 50% maximum optical density reading was reached is reported in Table 1. As compared to the two sham-operated controls, four of the six mice whose popliteal lymph node had been removed maintained comparable levels of specific antibody over the following 21 weeks. The remaining two mice gradually declined such that by 5 months their specific antibody titers were reduced to at least 50% of those in the sham-operated controls.

The differences between our results (i.e., only two out of six mice demonstrated the 50% reduction) and those reported by Tew et al. (1980) could be attributed to the fact that the latter assessed maintenance of titers associated with a secondary response, while we evaluated titers obtained after a single injection. The levels of specific antibody generated following a single immunization are generally much lower as compared to a second dose. In addition, the antigen in their study was emulsified in complete Freund's adjuvant, which also increases the titer of antibody relative to an alum-precipitated antigen. Physiologically, it may require less antigen to maintain the low amounts of antibody obtained in our study, while keeping over 400 μg/ml requires a higher amount of retained antigen. Thus, the small amount of antigen detected at 5

Table 1. Total antitetanus toxoid immunoglobulin titers

Mouse no.	1/Dilution of 50% maximum optical density[e]				
	21[c]/0[d]	22[c]/1[d]	23[3]/2[d]	32[c]/11[d]	42[c]/21[d]
Sham 1[a]	132	151	201	120	150
Sham 2[a]	138	97	81	151	100
Pop.1[b]	98	76	111	136	100
Pop.2[b]	120	85	107	120	120
Pop.3[b]	133	114	131	124	130
Pop.4[b]	96	76	99	71	<50
Pop.5[b]	199	233	295	235	150
Pop.6[b]	122	128	111	93	<50

[a] Sham 1 and 2 refer to the two control mice who were immunized once with tetanus in alum and underwent the surgical procedure without removing the draining popliteal lymph node 5 months after priming.

[b] Pop.1 – Pop.6 refer to the six experimental mice who were immunized once with tetanus in alum and underwent the surgical procedure to remove the draining popliteal lymph node 5 months after priming.

[c] Weeks after immunization with tetanus toxoid.

[d] Weeks aftger surgery to remove popiteal lymph node.

[e] As determined by a standard enzyme-linked immunosorbent assay (ELISA).

months in the para-aortic lymph node may have been enough to support the lower, but consistent levels, of specific antibody found in the mice in our experiments.

In both studies, the data indicate that the antigen retained in unremoved lymphoid tissue appears to be adequate for maintaining specific antibody titers. Part of the mechanism for this appears to be the local induction of preplasma cells only in germinal centers that contain the FDC-retained antigen (DONALDSON et al. 1986). Evidence for this was provided using an in vivo model in which immune mice were bled in order to remove half their volume of serum proteins. The sites at which the induction of new antigen-specific antibody cells (i.e., preplasma cells) occurred involved the lymph nodes where the antigen was localized.

Next, the relationship between retention of antigen on FDC and maintenance of memory cells was assessed. For this, an experimental protocol shown in Fig. 5A was used to measure the ability to make a secondary response after adoptive transfer of cells into severe combined immunodeficient (SCID) mice. The results (Fig. 5B) indicated that the group whose popliteal lymph node was removed 4 months previously (group 4) showed a comparable response to that observed in the immune sham-operated mice (group 3). In addition, these responses were comparable to mice receiving a booster immunization 2 weeks after surgery (group 6). Finally, similar antibody titers were achieved whether the cells used for transfer were isolated from the draining or nondraining lymph nodes or the spleen (data shown only for the draining set).

These experiments demonstrate that despite the removal of a sizeable amount of FDC-retained antigen, long-term antibody titers and circulating memory cells can be maintained by the small amount present in other lymph nodes. Maintenance of memory populations by cross-reactive antigens as recently discussed by MATZINGER (1994) may also contribute. However, the consistent finding of antigen in one of the para-aortic lymph nodes throughout our experiments supports the idea that antigen retained for very long times in seemingly small amounts is enough for the maintenance of the immune response.

3 Role of Follicular Dendritic Cells in B Cell Activation During Germinal Center Responses

The specific events leading to a germinal center reaction are not yet well defined. However, the response has been clearly shown to be dependent on not only FDC and B cells, but also on T cell help as well. Szakal, Tew, and colleagues described the cellular interactions that occur during the initial phase of the reaction (SZAKAL et al. 1989; TEW et al. 1989). This pathway involves the depositing of antigen on FDC and its subsequent uptake and processing by B cells (SZAKAL et al. 1988). During this process, the B cells acquire the ability to act as effective

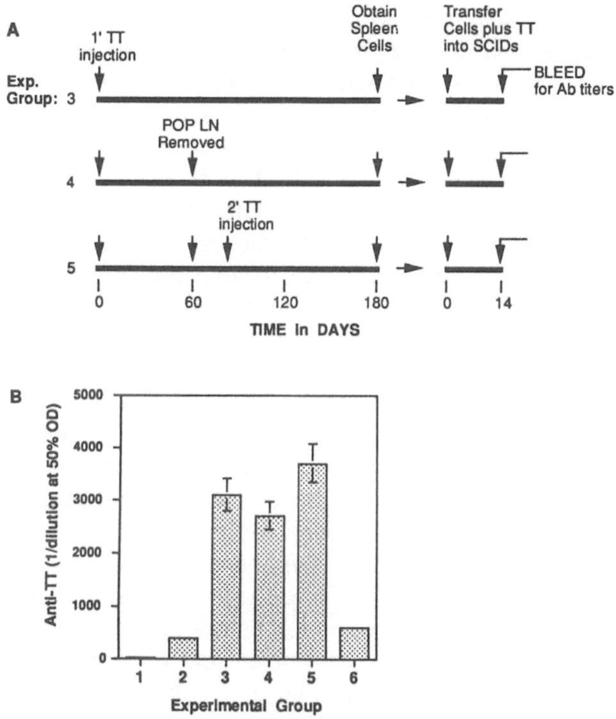

Fig. 5 A,B. The ability to generate a secondary immune response in adoptive severe combined immunodeficient (SCID) mouse recipients following removal of draining lymphoid organs from the donor Balb/c population. **A** Experimental protocol used to generate tetanus toxoid (*TT*)-immune mice (50 μg TT precipitated in alum, subcutaneously injected into left hind limb only) who were either sham operated or whose draining popliteal lymph node (*pop LN*) was removed. *Ab*, antibody. **B** Level of immunoglobulin G (IgG) anti-TT titers 2 weeks after adoptive transfer of 10^7 spleen cells (i.v.) and 50 μg soluble TT (intraperitoneally) into one of five experimental SCID mice groups. *Group 1* received only TT (no cells transferred); *group 2* received TT plus cells from naive donors; *group 3* received TT plus cells from sham-operated TT-immune mice (see A); *group 4* received TT plus cells from TT-immune mice whose draining popliteal lymph node was removed 2 months after primary immunization in their left hind limb only and 6 months prior to the adoptive transfer (see A); *group 5* received TT plus cells from TT-immune mice whose draining popliteal lymph node was removed 2 months after primary immunization in their left hind limb only and 6 months prior to the adoptive transfer, who also received a secondary immunization with soluble TT 1 week after surgery (see A); *group 6*, level of anti-TT titers in group 4 at the time the animals were killed to obtain their cells for the adoptive transfer. The values represent the reciprocal of the dilution obtained by enzyme-linked immunosorbent assay (ELISA) at which a 50% optical density (*OD*) reading was obtained over a dilution curve of the respective serum. Each group represents the titers of at least three mice ± the standard deviation value

antigen-presenting cells, consequently eliciting the appropriate T cell help (Kosco et al. 1988). Our experimental approach has been to generate an environment in vitro that mimics the characteristics of an in vivo germinal center. In this manner, we have been able to isolate individual components contributing to a germinal center reaction.

The initial studies involved displacing whole populations of germinal center cells from the tissue into single cell suspensions that formed clusters containing

FDC, B cells, and T cells during culture (Kosco et al. 1992). These experiments demonstrated that the adhesion/activation molecules intercellular adhesion molecule (ICAM)-1, lymphocyte function-associated antigen (LFA)-1, and CD44 were involved in cluster formation and B cell proliferation. We also attempted to block B cell proliferation or antibody production using combinations of neutralizing monoclonal antibodies directed against interleukin (IL)-1 to -6. However, using this strategy, any alteration from the control levels was difficult to achieve. This most probably was due to the close cell–cell contact, resulting in junctions that permitted the delivery of newly secreted cytokines to a receptor but excluded the exogenous antibodies present in the extracellular milieu. Anti-IL-4 was one exception in that adding this neutralizing antibody consistently lowered the amount of IgG_1 and IgE produced in these cultures (M. Kosco-Vilbois, G. LeGros, and D. Scheidegger, unpububLished data).

The second strategy pursued involved the isolation of individual cell types, i.e., FDC, B cells, and T cells, which were then added together for short-term cell culture assays. This facilitated the study of the initial events contributing to germinal center formation. For example, small, resting, recirculating B cells were obtained from transgenic mice and evaluated for the phenotypic changes that occurred following in vitro interactions with FDC (Kosco-Vilbois et al. 1993). These B cells, separated from activated B cells on the basis of buoyant densities, were found to be activated differentially by FDC in an antigen-independent versus-dependent manner. The independent signal involved an increase in size and complexity and also the level of major histocompatibility complex (MHC) class II expression. These parameters occurred whether or not the FDC carried immune complexes containing the antigen which could cross-link the B cell's immunoglobulin molecules. Possible candidates for eliciting these effects include the fragments of complement (e.g., C3b, C3d) present on the FDC's plasma membrane. It has been shown that receptors for these fragments expressed on B cells aid in the process of their activation (Carter and Fearon 1989; Matsumoto et al. 1991). Such possibilities are currently being explored.

In addition to these changes, FDC also brought about the de novo expression of B7 by the resting B cells. This effect is particularly significant, since it has been shown to be an essential second signal for T cell activation (Freedman et al. 1987; Linsley et al. 1990). In contrast to the other changes, though, the induction of B7 expression was absolutely dependent on having the appropriate antigen on the surface of the FDC for cross-linking the B cell's immunoglobulin molecules (Kosco-Vilbois et al. 1993).

The changes that occur after coincubation with FDC result in the capacity of B cells to present antigen. Thus, these experiments strongly suggest that in vivo small, recirculating B cells can encounter FDC and their retained antigen, become activated, and effectively present the antigen to T cells. The induction of higher levels of MHC class II and the costimulatory molecule B7 on potential germinal center B cells may be the earliest events required for producing a germinal center reaction.

In support of an early role for B7 in generating germinal center responses, an in vivo model was developed. For this, transgenic mice were generated that maintained in their circulation 10–30 μg/ml of a chimeric immunoglobulin molecule that would effectively block B7–ligand interactions (LANE et al. 1994; RONCHESE et al. 1994). Using the higher affinity of the two known ligands which bind to B7, i.e., CD28 and CTLA-4 (LINSLEY et al. 1990, 1992), a CTLA-4–Ig molecule was constructed of the extracellular portion of CTLA-4 (BRUNET et al. 1987) and the human IgG_1 constant domains. When $CD4^+$ T cell priming was assessed in these transgenic mice, no impairment was found (RONCHESE et al. 1994). However, antibody responses to T-dependent antigens were significantly affected (LANE et al. 1994). While antigen-specific IgM levels were comparable to wild-type controls, all IgG isotypes were reduced at least tenfold. In addition, germinal center formation was absent in these mice even after multiple immunizations with antigen. Using the antigen 4-hydroxy-3-nitrophenyl-acetyl chicken gamma globulin (NP-CγG), a reduction in cells undergoing somatic mutation and selection could be documented. These observations of minimal somatic mutation and selection as well as impaired class switching correlated with the lack of germinal centers. From this model, it appears that the B7 costimulatory signal is required for effective cellular interactions leading to germinal center reactions. In the absence of this signal, effective humoral responses cannot be elicited.

During our study of immune responses in CTLA-4–Ig transgenic mice, several peculiar observations in relation to FDC were made. First, although FDC trapped and retained the chimeric construct, no evidence was found of trapping antigen-containing immune complexes following a primary or secondary immunization of the transgenic mice (LANE et al. 1994). Second, immune complexes did not localize on FDC of transgenic mice following either passive immunization or using antibody aggregates in conjuction with immuno-histochemical techniques. This impairment could not be attributed to a lack of relevent receptors, since in situ FDC labeled with the monoclonal antibodies 2.4G2, 7E9, and 8C12. These reagents recognize the Fc receptor of IgG or complement receptors CR1 or CR2, respectively. Another observation was that the size of the FDC network was significantly smaller compared to wild-type controls. These observations provide additional evidence that FDC function is affected by impaired B and T cell interactions.

Unfortunately, to date the nature of these mechanisms involving B or T cell induction of FDC function has been only minimally addressed. This has mainly been due to the difficulty in obtaining purified populations of isolated FDC. In addition, FDC most likely exist in different stages of maturation, which must also be defined and studied (RADEMAKERS 1992). It is clear, though, from studies involving the de novo induction of an FDC network in SCID mice, that both B and T cells contribute to achieve full FDC competence (KAPASI et al. 1993). Recently conducted experiments which begin to address these questions are discussed in this volume by G. GROUARD et al., E. HEINEN et al., J. TEW et al., and T. VAN DEN BERG et al. New technologies such as transgenic mouse systems and the establishment of FDC lines such as those recently reported (ORSCHESCHEK et al. 1994; LINDHOUT et al. 1993) will also facilitate research into these areas.

4 Is Interleukin-6 Produced by Follicular Dendritic Cells?

Several laboratories have attempted to define the cytokines that are produced during germinal center reactions. The production of IL-6 has been of particular interest because of its influence on the terminal differentiation of B cells (KISHIMOTO 1989). The level of IL-6 gene expression by germinal center cells has been evaluated using two strategies providing varying results. In human tonsils obtained from children with recurrent throat infections, IL-6 gene expression detected by in situ hybridization was only found in extrafollicular areas (BOSSELOIR et al. 1989). Another study analyzed lymph nodes of patients suffering from Castleman's disease that contained numerous hyperplastic follicles (LEGER-RAVET et al. 1991). The FDC in these tissues were shown to express both IL-6 mRNA and protein as determined by in situ hybridization and immunohistochemistry. A recent study by ORSCHESCHEK and colleagues (1994) involving the generation of a human FDC-like cell line also addressed this issue. They reported that their immortalized FDC contained mRNA for IL-6 and further supported their findings by demonstrating IL-6 mRNA associated with FDC in frozen sections of human tonsils. However, two different laboratories have failed to detect IL-6 mRNA in FDC-enriched single cell preparations from human tonsils (BUTCH et al. 1993; SCHRIEVER et al. 1991).

In the above studies, the inability to detect IL-6 may reflect the fact that the regulation of the gene is induced only during short intervals of the germinal center reaction. In mice rendered deficient for IL-6 by gene targeting, antibody responses and germinal center development were impaired (KOSCO-VILBOIS et al., in preparation). While IgM levels were comparable to those in control mice when responding to a T cell-dependent antigen, IgG_1, IgG_{2a}, IgG_{2b}, and IgG_3 were significantly lower in the IL-6-deficient mice. Interestingly, IgG_1 levels with time increased such that eventually they were equal to the controls, while the other IgG isotypes remained low or absent. Histochemical evaluation of the germinal center reaction revealed that small foci of peanut agglutinin (PNA)-binding B cells and FDC form within the primary follicles. However, enlarged dark and light zones did not develop. Using PNA binding as a marker for germinal centers (COICO et al. 1983) in conjunction with morphometric analysis clearly documented a difference in size. While control mice produced maximum dimensions of 1.5–2.0 mm^3 between days 9 and 11 after primary immunization, at no time during the response was any individual PNA positive area in the IL-6-deficient mice larger than 0.4 mm^3. These results strongly suggest that IL-6 is necessary for providing the appropriate number of B cells that can contribute to the germinal center reaction. In the IL-6-deficient mice, the impairment in achieving normal levels of cells within a germinal center translated into lower antibody titers.

The relevance of IL-6 production in germinal centers to this review is shown by our data demonstrating that preparations containing FDC express both IL-6 mRNA and protein (M. Kosco-Vilbois, D. Scheidegger, and M. Wiles, unpublished observations). By reverse transcriptase polymerase chain reaction (RT-PCR), lymph node isolates containing up to 40% FDC produced a strong IL-6 signal, while relatively pure populations of germinal center B and T cells obtained by fluorescence-activated cell sorting (FACS) did not. While we are currently striving to demonstrate this in vivo by in situ hybridization on frozen sections, the FDC and the tingible body macrophage are the two remaining candidates as the source of the IL-6. In light of the studies performed with human tissue, the data supports the proposal that FDC produces IL-6 in particular situations. Verification of the cell producing IL-6 in germinal centers using the mouse system remains one of our major objectives.

Acknowledgments. We wish to thank Dr. I. Esparza, Dr. D. Gray, Dr. M. Julius, Dr. P. Lane, Dr. M. Wiles, Dr. Z. Kapasi, and Dr. M. Kopf, who collaborated on the various studies reported here and Dr. F. McConnell and Dr. I.-L. Martensson-Bopp for their critical reading of the manuscript. We would also like to gratefully acknowledge the excellent technical support of B. Pfeiffer, B. Kugelberg, and B. Johansson. The Basel Institute for Immunology was founded and is supported by F.Hoffmann-La Roche, Basel, Switzerland.

References

Bosseloir A, Hooghe-Peters EL, Heinen E, Cormann N, Kinet-Donoel C, Simar L (1989) Localization of interleukin 6 mRNA in human tonsils by in situ hybridization. Eur J Immunol 19: 2379–2381
Brunet JF, Denizot F, Luciani MF, Roux DM, Suzan M, Mattei MG, Golstein P (1987) A new member of the immunoglobulin superfamily—CTLA-4. Nature 328: 267–269
Butch AW, Chung GH, Hoffmann JW, Nahm MH (1993) Cytokine expression by germinal center cells. J Immunol 150: 39–47
Carter RH, Fearon DT (1989) Polymeric C3dg primes human B lymphocytes for proliferation induced by anti-IgM. J Immunol 143: 1755–1760
Coico RF, Bhogal S, Thorbecke GJ (1983) Relationship of germinal centers in lymphoid tissue to immunologic memory. VI. Transfer of B cell memory with lymph node cells fractionated according to their receptors for peanut agglutinin. J Immunol 131: 2254–2257
Donaldson SL, Kosco MH, Szakal AK, Tew JG (1986) Localization of antibody-forming cells in draining lymphoid organs during long-term maintenance of the antibody response. J Leukoc Biol 40: 147–157
Freedman AS, Freeman G, Horowitz JC, Daley J, Nadler LM (1987) B7, a B cell-restricted antigen that identifies preactivated B cells. J Immunol 139: 3260–3267
Gray D (1992) The dynamics of immunological memory. Semin Immunol 4: 29–34
Gray D, Leanderson T (1990) Expansion, selection and maintenance of memory B-cell clones. In: Gray D, Sprent I (eds) Immunological memory. Springer, Berlin Heidelberg New York, pp 1–17 (Current topics in microbiology and immunology, vol 159)
Gray D, Skarvall H (1988) B cell memory is short-lived in the absence of antigen. Nature 336: 70–73
Ho F, Lortan JE, MacLennan ICM, Khan M (1986) Distinct short-lived and long-lived antibody producing cell populations. Eur J Immunol 16: 1297–1301
Kapasi ZF, Burton GF, Shultz LD, Tew JG, Szakal AK (1993) Induction of functional follicular dendritic cell development in severe combined immunodeficiency mice. Influence of B and T cells. J Immunol. 150: 2648–2658
Kishimoto T (1989) The biology of interleukin-6. Blood 74: 1–10
Kosco MH, Gray D (1992) Signals involved in germinal center reactions. Immunol Rev 126: 63–76

Kosco MH, Szakal AK, Tew JG (1988) In vivo obtained antigen presented by germinal center B cells to T cells in vitro. J Immunol 140: 354–360

Kosco MH, Pflugfelder E, Gray D (1992) Follicular dendritic cell-dependent adhesion and proliferation of B cells in vitro. J Immunol 148: 2331–2339

Kosco-Vilbois MH, Gray D, Scheidegger D, Julius M (1993) Follicular dendritic cells help resting B cells to become effective antigen presenting cells: induction of B7/BB1 and upregulation of MHC class II molecules. J Exp Med 178: 2055–2066

Lane P, Burdet C, Hubele S, Scheidegger D, Müller U, McConnell F, Kosco-Vilbois M (1994) B cell function in mice transgenic for mCTKA4-Hγ1: lack of germinal centers correlated with poor affinity maturation and class switching despite normal priming of CD4+T cells. J Exp Med 179: 810–830

Leger-Ravet MB, Peuchmaur M, Devergen O, Audouin J, Raphael M, Van Damme J, Galanaud P, Diebold J, Emilie D (1991) Interleukin-6 gene expression in Castleman's disease. Blood 78: 2923–2930

Lindhout E, Mevissen ML, Kwekkeboom J, Tager JM, de Groot C (1993) Direct evidence that human follicular dendritic cells (FDC) rescue germinal centre B cells from death by apoptosis. Clin Exp Immunol 91: 330–336

Linsley PS, Clark EA, Ledbetter JA (1990) T cell antigen CD28 mediates adhesion with B cells by interacting with activation antigen B7/BB1. Proc Natl Acad Sci USA 87: 5031–5035

Linsley PS, Greene JL, Tan P, Bradshaw J, Ledbetter JA, Anasetti C, Damle NK (1992) Coexpression and functional cooperation of CTLA-4 and CD28 on activated T lymphocytes. J Exp Med 176: 1595–1604

MacLennan ICM (1994) Germinal centers. Annu Rev Immunol 12: 117–139

Mandel TE, Phipps RP, Abbot A, Tew JG (1980) The follicular dendritic cell: long term antigen retention during immunity. Immunol Rev 53: 29

Matsumoto AK, Kopicky-Burd J, Carter RH, Tuveson DA, Tedder TF, Fearon DT (1991) Intersection of the complement and immune systems: a signal transduction complex of the B lymphocyte-containing complement receptor type 2 and CD19. J Exp Med 173: 55–64

Matzinger P (1994) Memories are made of this? Nature 369: 605–606

Orscheschek K , Merz H, Schlegelberger B, Feller A (1994) An immortalized cell line with features of human follicular dendritic cells. Eur J Immunol 24: 2682–2690

Rademakers LH (1992) Dark and light zones of germinal centres of the human tonsil: an ultrastructural study with emphasis on heterogeneity of follicular dendritic cells. Cell Tissue Res 269: 359–368

Ronchese F, Housemann B, Hubele S, Lane P (1994) Mice transgenic for a soluble form of murine CTLA-4 show enhanced expansion of antigen-specific CD4+ T cells defective antibody production in vivo. J Exp Med 179: 809–817

Schriever F, Freeman G, Nadler LM (1991) Follicular dendritic cells contain a unique gene repertoire demonstrated by single-cell polymerase chain reaction. Blood 77: 787–791

Szakal AK, Kosco MH, Tew JG (1988) A novel in vivo follicular dendritic cell-dependent iccosome-mediated mechanism for delivery of antigen to antigen-processing cells. J Immunol 140: 341–353

Szakal AK, Kosco MH, Tew JG (1989) Microanatomy of lymphoid tissue during humoral immune responses: structure function relationships. Annu Rev Immunol 7: 91–110

Tew JG, Phipps RP, Mandel TE (1980) The maintenance and regulation of the humoral immune response: persisting antigen and the role of follicular antigen-binding dendritic cells as accessory cells. Immunol Rev 53: 175–201

Tew JG, Kosco MH, Szakal AK (1989) The alternative antigen pathway. Immunol Today 10: 229–232

Tew JG, DiLosa RM, Burton GF, Kosco MH, Kupp LI, Masuda A, Szakal AK (1992) Germinal centers and antibody production in bone marrow. Immunol Rev 126: 99–112

Role of Follicular Dendritic Cells in the Regulation of B Cell Proliferation

A.S. FREEDMAN, D. WANG, J. SCOTT PHIFER, and S.N. MANIE

1 Introduction

Within secondary lymphoid organs, B and T lymphocytes are not randomly distributed but they specifically inhabit highly organized regions that contain areas where B and T cells are largely segregated and other areas where different populations of cells can interact and differentiate (THORBECKE et al. 1962). Specifically, in lymph nodes and tonsils, the majority of B cells localize to follicles, whereas most T cells are found in the paracortex (NIEUWENHUIS and FORD 1976). Following initial entrance into the interfollicular areas of secondary lymphoid tissues, it is not known what drives B cells to localize to follicles; however, the formation of follicles is an antigen (Ag)-driven process that requires the presence and participation of T lymphocytes.

Primary follicles are largely composed of recirculating surface immunoglobulin (sIg) M/D$^+$ B cells, small numbers of CD4$^+$ T cells, and small numbers of follicular dendritic cells (FDC) (MACLENNAN et al. 1990; STEIN et al. 1982). B cell blasts from the interfollicular areas migrate into primary follicles and initiate the formation of germinal centers (GC). It has been proposed that the function of the GC is to generate cells that are destined to secrete immunoglobulin (Ig), the site of Ig heavy chain class switch, the production of memory B cells, and the maturation of antibody affinity (KLAUS et al. 1980; MACLENNAN et al. 1990; NIEUWENHUIS and KEUNING 1974; THORBECKE et al. 1962). During early GC formation, B cells are enveloped in a network of FDC. B cell blasts (centroblasts) occupy the

Division of Hematologic Malignancies, Dana-Farber Cancer Institute, Department of Medicine, Harvard Medical School, Boston, MA 02115, USA basal zone (dark zone) of the follicle, in which there are few associated FDC (less than 2% of the cell population). Another population of noncycling B cells develop at the opposite pole of the GC, referred to as the light zone. The B cells in this area, termed centrocytes, are associated with a dense network of FDC (10%–20% of the GC volume). The secondary follicle later evolves to a structure in which there are small numbers of centroblasts associated with FDC.

In normal GC, FDC play a pivotal role in the development of secondary immune responses (KLAUS et al. 1980; TEW et al. 1990). FDC are often most prominent in the light zone area of the GC. Developmentally, B cells in the light zone are cells at the junction of apoptotic cell death or differentiation into either memory B cells or antibody-secreting cells. Although Ag presentation by FDC to B cells is critical in this process, it is likely that cell–cell contact signals and/or cytokines are also important in B cell–FDC interactions. Recent studies have provided insight into the nature of the cell–cell interactions between B cells and FDC which may regulate B cell proliferation and the differentiation. In this review we will detail the nature of the adhesive interactions of FDC and B cells and the functional consequences of those interactions.

2 B Cell–Follicular Dendritic Cell Adhesive Interactions

The expression of cell surface Ag by FDC reflect their proposed function in Ag presentation to B cells and cell–cell adhesion (JOHNSON et al. 1986; PETRASCHET et al. 1990; SCHNIZLEIN et al. 1985; SCHRIEVER et al. 1989; SELLHEYER et al. 1989; SZAKAL et al. 1985). FDC present Ag to B cells in the form of immune complexes and have the capacity to retain these complexes on their surfaces for periods of weeks to months. Consistent with this ability to fix and express immune complexes, FDC express complement cleavage fragment (C3b, C3bi, and C3d) and Fc (IgE) receptors. The expression of cell adhesion molecules on FDC, e.g. intercellular adhesion molecule (ICAM)-1 and vascular cell adhesion molecule (VCAM)-1, reflect their role in cell–cell interactions through interaction with their respective ligands on lymphocytes, lymphocyte function-associated antigen (LFA-1) and very late antigen (VLA)-4.

Insight into the adhesive interactions between FDC and B cells have come from in vitro adhesion assays. Several years ago we developed a tissue-binding assay to examine B cell–FDC interactions (FREEDMAN et al. 1990). This technique involved the use of normal splenic or tonsillar B cells cultured for 3 days with *Staphylococcus aureus* Cowan I (SAC) internally labeled with a fluorochrome, and then incubated at 25°C on rotated frozen sections of human tonsil. These cells adhered preferentially to GC, as demonstrated by combined fluorescence and brightfield microscopy. The distribution of cells adhering to GC was identical to the pattern of staining of a parallel section of tonsil with a monoclonal antibody (mAb) specific for FDC. Therefore it appeared that the applied cells were binding to FDC. In contrast, unstimulated B cells demonstrated little to no binding to GC at 25°C. The binding of SAG-stimulated cells to the GC was essentially absent 4°C.

To identify the receptor–ligand pair(s) involved in B cell–GC adhesion, we examined several lymphoid cell lines with distinct cell surface phenotypic and GC-binding characteristics. We found that there was no clear relationship between GC binding and expression of known adhesion receptors. However, the B cell line Nalm-6 had a high-affinity binding to GC. The cell surface phenotype of Nalm-6 cells was studied and we performed blocking studies with all positively reacting mAb to determine which, if any, of the cell surface Ag expressed by Nalm-6 were involved in GC binding. The adhesion of Nalm-6 to GC was blocked by anti-CD29 and anti-CD49d (VLA-4, $\alpha_4\beta_1$-integrin). Of the β_1-integrins, VLA-4 is unique since it can function as a cell–extracellular matrix and cell–cell adhesion receptor. VLA-4 has two ligands, the FN-40 fragment of fibronectin and VCAM-1, a cell surface protein originally identified on cytokine-activated endothelium (ELICES et al. 1990; GUAN and HYNES 1990) that we knew from prior immunohistochemistry studies was also densely expressed on FDC. Using the frozen section binding assay, mAb against VCAM-1 blocked Nalm-6 binding by nearly 100%. Therefore, B lymphocyte binding to GC was the result of a direct interaction betwen VLA-4 on lymphoid cells and VCAM-1 expressed on FDC (FREEDMAN et al. 1990). These results have been confirmed and extended by Koopman et al. using isolated FDC and B cells from tonsil. By examining cluster formation of aggregates of B cells and FDC, it was found that mAb against VLA-4 as well as LFA-1 on the B cells and VCAM-1 and ICAM-1 on the FDC inhibited B cell–FDC adhesion (KOOPMAN et al. 1991). This provided evidence for both β_1- and β_2-integrin-mediated pathways in B cell adhesion to FDC. These results have been confirmed in murine systems, where FDC–B cell cluster formation was inhibited by mAb directed agianst ICAM-1, LFA-1, and anti-kappa mAb (Kosco et al. 1992). The adhesion molecule CD44 did not have effects in either human or murine systems on cluster formation. These studies are important to the regualtion of B cell proliferation, since FDC, in addition to presenting antigen, can regulate B cell activation through antigen-independent cell–cell interactions.

3 Follicular Dendritic Cell–B Cell Costimulation

Based upon the in situ observations that FDC are present in GC, it has long been hypothesized that FDC provide costimulatory signals to B cells. In studies by Schnizlein and coworkers in both murine and human systems using only partially purified preparations of FDC, the addition of those cells to polyclonally activated B cells augmented B cell proliferation (SCHNIZLEIN et al. 1984). Further evidence for the growth-enhancing effect of FDC upon GC B cells stems from a study using cell suspensions. In these studies approximately 7% of cells were FDC and over

90% had the surface phenotype of GC B cells ($CD19^+/CD29^-/IgD^-$). The percentage of cells clustered with the FDC staining positively with Ki67, a marker of activated proliferating cells, was significantly higher than the cells outside of the clusters. There was also evidence of more apoptotic cells outside of the FDC–B cell clusters than within (PETRASCH et al. 1991). More recently, techniques have been developed to enrich cell suspensions of lymphoid tissues for FDC to near homogeneity. Using more highly purified FDC, insight into the effects of FDC on B cell proliferation has become more clear.

There is a body of evidence that GC (low-density) B cell proliferation is dependent on the presence of FDC. KOSCO et al. (1992) demonstrated that clusters formed in vitro of FDC, T cells, and B cells contain proliferating B cells. The formation of clusters is FDC and, to a lesser degree, T cell dependent. More importantly, depletion of T cells and FDC led to a marked decrease in proliferation of B cells. This FDC-dependent proliferation is highest during the first 4 days of culture, declining by days 7–11. In studies by BURTON et al. (1993), enriched FDC enhanced the proliferation of murine B cells as measured by thymidine incorporation. This was observed with B cells activated through the Ag receptor, with anti-μ, as well as with lipopolysaccharide (LPS) activation. Little to no B cell proliferation was seen in the absence of the polyclonal B cell stimuli. The effect of FDC was maximal at 1–2 days of culture and declined thereafter, but persisted at 4 days. These cultures were also examined for cell viability, and it was observed that wells containing FDC contained significantly greater numbers of viable B cells than wells containing only B cells. Of interest, the enhancing effect of FDC did not require the presence of T cells. Similar findings have been reported using cultured FDC (CLARK et al. 1992). In those studies, purified FDC which had been maintained in culture with granulocyte-macrophage colony-stimulating factor (GM-CSF) were able to augment the proliferation of human tonsillar B cells which were costimulated with anti-human Ig, anti-CD40 mAb, and a combination of both. These studies support the role of FDC as costimulatory cells for B cells proliferation, recapitulating the in vivo finding of B cell proliferation in GC.

In contrast to these reports, our laboratory observed that purified human FDC can inhibit B cell proliferation. We isolated homogeneous preparations of FDC from normal tonsil to begin to examine their effects in vitro on B cell proliferation (FREEDMAN et al. 1992). These FDC were isolated from the low-density fraction (35% interface of a Percoll gradient) of tonsil cell suspensions. FDC were then further enriched by cell sorting using the FDC-restricted mAb DRC-1 (> 80% FDC). When irradiated FDC were cultured with mitogen-stimulated B cells, B cell [^{3}H]thymidine uptake was inhibited by up to 80%. This was seen when FDC were added to previously activated B cells or when cocultured with mitogen and unactivated B cells. This inhibitory effect was not seen when paraformaldehyde-fixed FDC were added to B cell cultures, suggesting that the FDC needed to be metabolically active. Furthermore, supernatants from cultured FDC were able to partially inhibit B cell proliferation. This suggested that cell–cell contact signals which can be preserved by paraformaldehyde fixation may not be responsible for the observed effect. Alternatively, the critical cell–cell contact signals involved may in fact be destroyed by fixation. Our result may also reflect the activation state of the B cells or the FDC, as well as the methods of isolation of the cell populations. These findings have been also observed by LINDHOUT et al. (1994), who found that freshly isolated FDC inhibited proliferation of tonsillar B cells which had been activated with anti-IgM or the combination of SAC plus interleukin (IL)-2. The same effect on B cell proliferation was noted using FDC which had been transformed with Epstein-Barr virus (EBV). Moreover, the use of supernatants from the EBV-transformed FDC had inhibitory effects on polyclonal-activated B cell proliferation. These studies support the notion that the inhibition of proliferation may be an in vitro demonstration of the downregulation of proliferation which occurs in the light zone of GC.

4 B Cell Apoptosis and Follicular Dendritic Cells

In addition to serving as areas of intense B cell proliferation, GC are also sites of B cell death. This has been substantiated by the findings that GC B cells, when isolated, undergo apoptosis in culture. This apoptosis can be prevented by a number of exogenous stimuli, including combinations of anti-Ig and anti-CD40 mAb, as well as the combination of IL-1 and a soluble form of the CD23 molecule (LIU et al. 1989, 1991). Several investigators have reported that GC B Cells, when clustered with FDC, remain viable, while B cells outsie of the clusters die by apoptosis (KOSCO et al. 1992; LINDHOUT et al. 1993, 1994; KOOPMAN et al. 1994). The prevention of apoptosis had been shown to be dependent upon cell–cell adhesion between B cells and FDC, since mAb which inhibit those interactions, specifically directed against CD49d–CD106 (VLA-4–VCAM-1) and CD11a/18–CD54(LFA-1–ICAM-1), leads to apoptosis of the B cells (KOSCO et al. 1992; LINDHOUT et al. 1993; KOOPMAN et al. 1994). Moreover, there is evidence for the hypothesis that signaling through these adhesion molecules is critical to the prevention of apoptosis. Soluble forms of CD54 (ICAM-1) and CD106 (VCAM-1) were able to partially prevent apoptosis of GC B cells (KOOPMAN et al. 1994). This effect was enhanced by the presence of anti-IgM. Other studies have shown that, although they do not affect B cell–FDC adhesion, mAb against CD 44 and major histocompatibility complex (MHC) class II inhibited FDC augmentation of B cell proliferation (Kosco et al. 1992).

5 Signaling to B Cells Through Very Late Antigen-4

Recent studies have demonstrated that integrins are involved in the transduction of signals into cells (HYNES 1992). The mechanisms by which integrins initiate signal cascades are unknown, but they probably employ intermediary signaling proteins, since integrins themselves contain no identified intrinsic enzymatic activity. Cross-linking of integrins on a variety of cell types with mAb or natural ligands induces tyrosine phosphorylation of major substrates of 115–130 kDa (FERRELL and MARTIN 1989; GUAN et al. 1991; HAMAWY et al. 1993; HUANG et al. 1993; HYNES 1992; KORNBERG et al. 1991; NOJIMA et al. 1992; SCHALLER et al. 1992). One of these Tyr-phosphorylated proteins is a protein Tyr kinase (pp 125^{FAK}) which has been shown to localize to cellular focal adhesion areas in chicken embryo cells (SCHALLER et al. 1992). With the evidence that signaling through integrins is critical to cell growth and survival (FRISCH and FRANCIS 1994; MEREDITH et al. 1993; MORTARINI et al. 1992; SUGAHARA et al. 1994), studies of signaling through VLA-4 may provide important insights into the interactions of normal and neoplastic B cells with ligands present in their microenvironment (JULIANO and VARNER 1993; SCHWARTZ 1993).

A critical step towards understanding the role of VLA-4-mediated adhesion in B cell function would be to examine B cell stimulation through VLA-4. We have examined the effect of ligation of VLA-4 on protein Tyr phosphorylation (FREEDMAN et al. 1993; Manie et al., unpublished). Ligation of VLA-4 on normal B cells and B cell lines (Nalm-6, SB) by cross-linking with mAb or natural ligands induces Tyr phosphorylation of proteins of 110–130kDa, as determined by anti-phosphotyrosine immunoblotting (Fig 1). The 110-kDa protein pp 110 is the major species observed which undergoes increased Tyr phosphorylation. The increased Tyr phosphorylation of this protein was also seen in an EBV-transformed B cell line (SB) and normal tonsillar B cells. An increase in Tyr phosphorylation of pp 110 was seen 1 min after cross-linking of VLA-4, was maximal after 5–10 min, and persisted at 60 min. pp110 protein was not the β_1-chain of VLA-4, as shown by immunoprecipitation studies. In an attempt to identify the nature of pp110 we have determined by western blotting and/or immunoprecipitation that it is not vinculin (120 kDa), α-actinin (100 kDa), phosphatidylinositol (PI)-3 kinase (110-kDa subunit), AFAP-110 (actin filament-associated protein, $pp60^{src}$ substrate), $p120^{GAP}$, JAK - 3 (110 kDa), or the focal adhesion-associated Tyr kinase $pp125^{FAK}$.

We identified one of the 120- to 130-kDa proteins as $pp125^{FAK}$, as detected by immunoprecipitation following by western blot using anti-$pp125^{FAK}$ mAb. The increased phosphorylation of $pp125^{FAK}$ in vivo has been correlated with an increase in its kinase activity in vitro (GUAN and SHALLOWAY 1992). Furthermore, we have observed that $pp125^{FAK}$ isolated from VLA-4-activated Nalm 6 cells demonstrate augmented Tyr phosphorylation in in vitro immune complex kinase assays (Fig. 2). This is likely to be due to the autophosphorylation kinase activity of $pp125^{FAK}$ (HILDEBRAND et al. 1993). Thus $pp125^{FAK}$ is different from pp110 and, in Nalm-6, $pp125^{FAK}$ is one of the Tyr kinases whose phosphorylation is stimulated by VLA-4 ligation.

Several lines of evidence suggest that integrin-stimulated Tyr kinases are associated with the integrin-induced organization of the cytoskeleton. One of the most relevant observation is that cytochalasin D pretreatment, which prevents elongation of newly formed F-actin and disrupts actin filament organization, can block integrin stimulation of the Tyr phosphorylation of $pp125^{FAK}$ and other

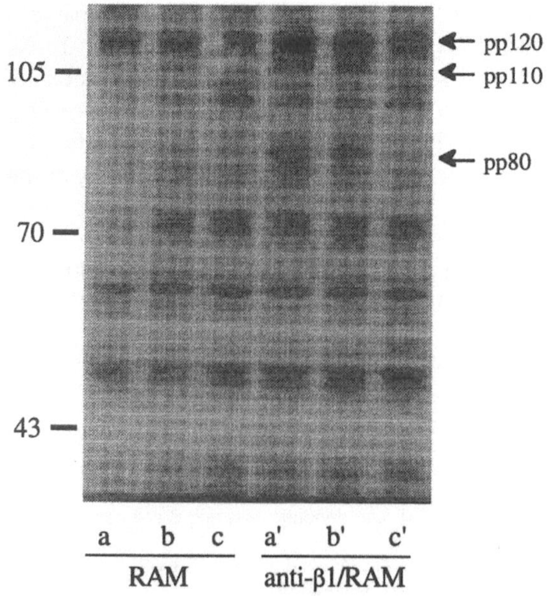

Fig. 1. β_1 Integrin ligation induces Tyr phosphorylation of multiple protein subtrates. Nalm-6 cells were preincubated for 1 h at 37°C with media alone (*a*), 1 μ*M* nocodazole (*b*), or 1 μ*M* cytochalasin D (*c*). Cells were then incubated for 20 min at 4°C with media alone (*a–c*) or anti-β_1-integrin monoclonal antibody (mAb) (*a'–b'*), followed by a 15-min stimulation at 37°C in the presence of rabbit anti-mouse antibodies (*RAM*). NP-40 cell lysates were resolved by sodium dodecyl sulfate polyacrylamide gel electrophoresis (SDS-PAGE), transferred to nitrocellulose, and probed with anti-phosphotyrosine mAb (4G10) which was detected by the ECL system (Amersham, USA)

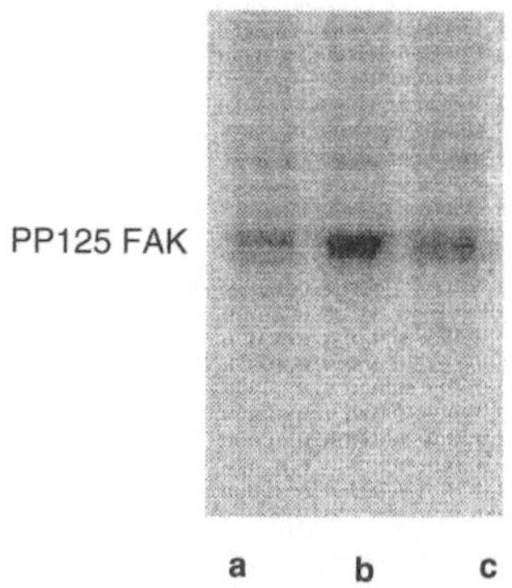

Fig. 2. β_1-Integrin ligation stimulates Tyr phosphorylation of pp 125^{FAK} in in vitro immune complex kinase assays. NP-40 lysates from unstimulated Nalm-6 cells (*a*) and Nalm-6 cells stimulated by ligation of β_1-integrin (*b*) or ligation of CD19 (*c*) were prepared. Cells lysates were then immunoprecipitated with anti-pp 125^{FAK} monoclonal antibody (mAb) (1F7). In vitro kinase assays using [γ^{32}P] ATP (adenosine triphosphate) were performed on immune complexes and subjected to sodium dodecyl sulfate polyacrylamide gel electrophoresis (SDS-PAGE). After alkali treatment, the gel was dried and autoradiographed

proteins (HUANG et al. 1993). We examined Tyr phosphorylation of pp110 and $pp125^{FAK}$ in Nalm-6 cells treated with cytochalasin D prior to VLA-4 cross-linking. The enhanced Tyr phosphorylation of both $pp125^{FAK}$ and pp110 following β_1-integrin cross-linking (Fig.1, lane a') is completely inhibited by pretreatment of cells with cytochalasin D (Fig.1, lane c'). In contrast, pretreatment of cells with nocodazole (which depolymerizes the microtubule network) is only partially effective in inhibiting VLA-4-induced Tyr phosphorylation (Fig.1, lane b'). These results suggest that folowing VLA-4 cross-linking on B cells, the reorganization of the cytoskeleton appears to function as a foundation for the building of a signaling transduction complex. Moreover, $pp125^{FAK}$ and pp110 are components of a common signal transduction pathway which is dependent on the integrity of the action cytoskeleton.

Acknowledgments. This work was supported by the National Institutes of Health (NIH) grant no. CA55207. SNM is supported by the Association pour la Recherche sur le Cancer, and the Hairy Cell Leukemia Foundation.

References

Burton GF, Conrad DH, Szakal AK, Tew JG (1993) Follicular dendritic cells and B cell costimulation. J Immunol 150: 31–38

Clark EA, Grabstein KH, Shu GL (1992) Cultured human follicular dendritic cells. Growth characteristics and interactions with B lymphocytes. J Immunol 148: 3327–3335

Elices MJ, Osborn L, Takada Y, Crouse C, Luhowskyj S, Hemler ME, Lobb RR (1990) VCAM-1 on activated endothelium interacts with the leukocyte integrin VLA-4, a site distinct from the VLA-4/fibronectin binding site. Cell 60: 577–584

Ferrell J, Martin G (1989) Tyrosine-specific protein phosphorylation is regulated by glycoprotein IIb-IIIa in platelets. Proc Natl Acad Sci USA 86: 2234–2238

Freedman As, Munro MJ, Rice GE, Bevilacqua MP, Morimoto C, McIntyre BW, Rhynhart K, Pober JS, Nadler LM (1990) Adhesion of human B cells to germinal centers in vitro involves VLA-4 and INCAM-110. Science 249: 1030–1033

Freedman A, Munro J, Rhynhart K, Schow P, Daley J, Lee N, Svahn J, Nadler L (1992) Follicular dendritic cells inhibit human B lymphocyte proliferation. Blood 80: 1284–1288

Freedman A, Rhynhart K, Nojima Y, Svahn J, Eliseo L, Benjamin C, Morimoto C, Vivier E (1993) Stimulation of protein tyrosine phosphorylation in human B cells following ligation of the β_1 integrin VLA-4. J Immunol 150: 1645–1652

Frisch S, Francis H (1994) Disruption of epithelial cell-matrix interactions induces apoptosis. J Cell Biol 124: 619–626

Guan JL, Hynes RO (1990) Lymphoid cells recognize an alternatively spliced segment of fibronectin via the integrin receptor $\alpha_4\beta_1$. Cell 60: 53–61

Guan JL, Shalloway D (1992) Regulation of focal adhesion-associated protein tyrosine kinase by both cellular adhesion and oncogenic transformation. Nature 358: 609–692

Guan J, Trevithick J, Hynes R (1991) Fibronectin/integrin interaction induces tyrosine phosphorylation of a 120-kDa protein. Cell Regul 2: 951–964

Hamawy M, Mergenhagen S, Siraganian R (1993) Tyrosine phosphorylation of $pp125^{FAK}$ by the aggregation of high affinity immunoglobin E receptors requires cell adherence. Biol Chem 268(10): 6851–6854

Hildebrand J, Schaller M, Parsons JT (1993) Identification of sequences required for the efficient localization of the focal adhesion kinase, $pp125^{FAK}$, to cellular focal adhesions. J Cell Biol 123: 993–1005

Huang M, Lipfert L, Cunningham M, Brugge J, Ginsberg M, Shattil S (1993) Adhesive ligand binding to integrin αIIbB3 stimulates tyrosine phosphorylation of novel protein subsrates before phosphorylation of $pp125^{FAK}$. J Cell Biol

Hynes R (1992) Integrins: Versatility, modulation, and signalling in cell adhesion. Cell 69: 11–25

Johnson GD, Hardie DL, Ling NR, MacLennan ICM (1986) Human follicular dendritic cells: a study with monoclonal antibodies. Clin Exp Immunol 64: 205–213

Juliano R, Varner J (1993) Adhesion molecules in cancer: the role of integrins. Curr Opin Cell Biol 5: 812–818
Klaus GGB, Humphrey JH, Kunkl A, Dongworth DW (1980) The follicular dendritic cell: its role in antigen presentation in the generation of immunological memory. In: Moeller G (ed) Immunological Reviews. Munksgaard, Copenhagen, pp 3–28
Koopman G, Parmentier HK, Schuurman H-K, Newman W, Meijer CJLM, Pals S (1991) Adhesion of human B cells to follicular dendritic cells involves both the lymphocyte function-associated antigen 1/intercellular adhesion molecule 1 and very late antigen 4/vascular cell adhesion molecule 1 pathways. J Exp Med 173: 1297–1304
Koopman G, Keehnen R, Lindhout E, Newman W, Shimuzu Y, Van Seventer G, De Groot C, Pals S (1994) Adhesion through the LFA-1 (CD11a/CD18)-ICAM-1(CD54) and the VLA-4 (CD49d)-VCAM-1 (CD106) pathways prevents apoptosis of germinal center B cells. J Immunol 152: 3760–3767
Kornberg L, Earp H,Turner C, Prockop C, Juliano R (1991) Signal transduction by integrins: increased protein tyrosine phosphorylation caused by clustering of β_1 integrins. Proc Natl Acad Sci USA 88: 8392–8396
Kosco MH, Pflugfelder E, Gray D (1992) Follicular dendritic cell-dependent adhesion and proliferation of B cells in vitro. J Immunol 148: 2331–2339
Lindhout E, Mevissen M, Kwekkeboom J, Tager J, DeGroot C (1993) Direct evidence that human folliculat dendritic cells (FDC) rescue germinal centre B cells from death by apoptosis. Clin Exp Immunol 91: 330–336
Lindhout E, Lakeman A, Mevissen M, DeGroot C (1994) Functionally active Epstein-Barr virus-transformed follicular dendritic cell-like lines. J Exp Med 179: 1173–1184
Liu Y, Cairns J, Holder M, Abbot S, Jansen K, Bonnefoy J, Gordoen J, MacLennan I (1991) Recombinant 25-kDa CD23 and interleukin 1 α promote the survival of germinal center B cells: evidence for bifurcation in the development of centrocytes rescued from apoptosis. Eur J Immunol 21: 1107–1114
Liu YJ, Joshua DE, Williams GT, Smith CA, Gordon J, MacLennan ICM (1989) Mechanism of antigen-driven selection in germinal centres. Nature 342: 929–931
MacLennan ICM, Liu YJ, Oldfield S, Zhang J, Lane PJL (1990) The evolution of B-cell clones. In: Gray D, Sprent J (eds) Immunological Memory. Springer, Berlin Heidelberg New York, pp 38–63
Meredith J, Fazeli B, Schwartz M (1993) The extracellular matrix as a cell survival factor. Mol Biol Cell 4: 953–961
Mortarini R, Gismondi A, Santoni A, Parmiani G, Anichini A (1992) Role of the $\alpha_5\beta_1$ integrin receptor in the proliferative response of quiescent human melanoma cells to fibronectin. Cancer Res 52: 4499–4506
Nieuwenhuis P, Ford WL (1976) Comparative migration of B and T lymphocytes in the rat spleen and lymph nodes. Cell Immunol 23: 254–267
Nieuwenhuis P, Keuning FJ (1974) Germinal centres and the origin of the B cell system II. Germinal centers in the rabbit spleen and popliteal lymph nodes. Immunology 26: 509–519
Nojima Y, Rothstein D, Sugita K, Schlossman S, Morimoto C (1992) Ligation of VLA-4 on T cells stimulated tyrosine phosphorylation of a 105-kD protein. J Exp Med 175: 1045–1053
Petrasch S, Perez-Alvarez C, Schmitz J, Kosco M, Brittinger G (1990) Antigenic phenotyping of human follicular dendritic cells isolated from nonmalignant and malignant lymphatic tissue. Eur J Immunol 20: 1013–1018
Petrasch SG, Kosco MH, Perez-Alvarez CJ, Schmitz J, Brittinger G (1991) Proliferation of germinal center B lymphocytes in vitro by direct membrane contact with follicular dendritic cells. Immunobiology 183: 451–462
Schaller M, Borgman C, Cobb B, Vines R, Reynolds A, Parsons JT (1992) $pp125^{FAK}$, a structurally distinctive protein-tyrosine kinase associated with focal adhesion. Proc Natl Acad Sci USA 89: 5192–5196
Schnizlein C, Szakal A, Tew J (1984) Follicular dendritic cells in the regulation and maintenance of immune responses. Immunobiology 168: 391
Schnizlein CT, Kosco MH, Szakal AK, Tew JG (1985) Follicular dendritic cells in suspension: identification, enrichment, and initial characterization indicating immune complex trapping and lack of adherence and phagocytic activity. J Immunol 134: 1360–1368
Schriever F, Freedman AS, Freeman G, Messner E, Lee G, Daley J, Nadler LM (1989) Isolated human follicular dendritic cells display a unique antigenic phenotype. J Exp Med 169: 2043–2058
Schwartz M (1993) Signaling by integrins: implications for tumorigenesis. Cancer Res 53: 1503–1506

Sellheyer K, Schwartinger R, Stein H (1989) Isolation and antigenic profile of follicular dendritic cells. Clin Exp Immunol 78: 431–436

Stein H, Gerdes J, Mason DY (1982) The normal and malignant germinal centre. Clin Haematol 11: 531–559

Sugahara H, Hanakura Y, Furitsu T, Ishihara K, Oritani K, Ikeda H, Kitayama H, Ishikawa J, Hashimoto K, Kanayama Y, Matsuzawa Y (1994) Induction of programmed cell death in human hematopoietic cell lines by fibronectin via its interaction with very late antigen 5. J Exp Med 179: 1757–1766

Szakal AK, Gieringer RL, Kosco MH, Tew JG (1985) Isolated follicular dendritic cells: cytochemical antigen localization, Normarski, SEM and TEM morphology. J Immunol 134: 1349–1353

Tew J, Kosco M, Burton G, Szakal A (1990) Follicular dendritic cells as accessory cells. Immunol Rev 117: 185–211

Thorbecke GJ, Asofsky R, Hochwald GM, Siskind GW (1962) Gamma globulin and antibody formation. J Exp Med 116: 295–310

Murine Follicular Dendritic Cells: Accessory Activities In Vitro

J.G. Tew[1], G.F. Burton[1], S. Helm[1], J. Wu[1], D. Qin[1], E. Hahn[1], and A.K. Szakal[2]

1 Introduction

Follicular dendritic cells (FDC) are accessory cells found in association with B lymphocytes in the follicles of stimulated secondary lymphoid tissues. Accessory cells are known to collaborate with lymphocytes and enable them to carry out their essential functions. FDC present antigen (Ag) to B cells and facilitate B cell functions; they may be thought of as B cell-associated accessory cells. In contrast, other dendritic cells, including Langerhans cells, interdigitating cells, and veiled cells, interact primarily with T lymphocytes. These dendritic accessory cells, collectively referred to as dendritic cells (DC) in vitro, present antigen to T cells and may be thought of as T cell-associated accessory cells (Tew et al. 1992a). Data supporting the accessory activities of most accessory cells, such as DC and macrophages, have been collected using in vitro experimentation. In contrast, most of the data available on FDC has been collected using in vivo models. We have reviewed these in vivo data on murine FDC in a number of recent articles.

[1]Department of Microbiology/Immunology, Medical College of Virginia, Virginia Commonwealth University, P.O. Box 980709, Richmond, VA 23298, USA
[2]Anatomy Division of Immunobiology, Medical College of Virginia, Virginia Commonwealth University, P.O. Box 980678, Richmond, VA 23298, USA

The major topics and references to these reviews are listed in Table 1. Primary data relevant to items C and E in Table 1 was recently published in a report by KAPASI et al. (1993) describing cellular requirements for reconstitution of FDC in severe combined immunodeficient (SCID) mice. Similarly, a recent data set relevant to item G is provided in a report by MASUDA et al. (1994) describing FDC function in murine acquired immunodeficiency syndrome (MAIDS). In view of these recent articles, we will concentrate on the recent in vitro data; we recommend the articles in Table 1 to the reader interested in the in vivo data.

The Ag retained on FDC is believed to interact with and stimulate B cells through specific immunoglobulin (Ig) receptors and thereby function in the induction and maintenance of humoral immune responses. In addition to specific Ag, FDC have other associated molecules which appear to be either costimulatory or are apparently capable of delivering activation signals to B cells directly (BURTON et al. 1993; KOSCO-VILBOIS et al. 1993). Methods for working with FDC and the FDC-derived immune complex-coated bodies (iccosomes) in vitro have been developed relatively recently (HUMPHREY and GRENNAN 1982; LILET-LECLERCQ et al. 1984; SCHNIZLEIN et al. 1985), including very recent modifications (KOSCO et al. 1992; BURTON et al. 1993). Some of this in vitro data is new, and numerous unpublished observations will be cited. The fact that these data are new and in some cases preliminary should be borne in mind by the reader.

Compared with other accessory cells, the data available on FDC functions are limited. In our judgement this is not because FDC are not important; rather, we attribute this lack of information to the fact that FDC are rare, fragile, and challenging to work with in vitro. However, we believe the data reviewed here support the hypothesis that FDC are critical for B lymphocytes to function fully

Table 1. Topics covered in recent reviews with emphasis on mirune Follicular Dendritic Cells (FDC)

Major topics reviewed	Review references
A. Cardinal features which distinguish FDC from all other cells	TEW et al. (1990, 1991, 1992a)
B. FDC ultrastructure and iccosome formation	SZAKAL et al. (1989) TEW et al. (1989)
C. "The alternative Ag pathway": Ag transport to FDC, iccosome formation, endocytosis, and processing of Ag by B cells	TEW et al. (1989m 1990, 1991) SZAKAL et al. (1992) SZAKAL TEW et al. (1992)
D. FDC involved in production of Ab and B memory cells: first Ab-forming cells and then B memory cells are produced	TEW et al. (1992b) BURTON et al. (1994)
E. FDC accessory activities and germinal center formation	TEW et al. (1990)
F. Potential of Ag targeted to FDC as a vaccine strategy	SZAKAL et al. (1991a) BURTON et al. (1994)
G. Effects of murine retroviruses on FDC and FDC function	SZAKAL et al. (1991b, 1992)
H. Methodology for isolating and working with FDC in vitro	TEW et al. (1994)
I. Origin of FDC	SZAKAL et al. (this volume)

Ag, antigen; Ab antibody

and appropriately. In short, we submit that an appreciation of the role of FDC in B cell functions is important to a full understanding of humoral immunity.

2 Accessory Activities of Follicular Dendritic Cells In Vitro

The accessory activities of FDC in vitro are listed in Table 2. Parts A and B describe the effects of FDC on B cells given a primary signal by mitogens. In this situation the B cells cluster around the FDC, and the FDC provide a costimulatory signal or signals which amplify the primary signal from the mitogen. The FDC alone are inactive or only poorly active in driving proliferation or eliciting antibody production by themselves. Parts C and D of Table 2 represent the effect of Ag-bearing FDC on proliferation and antibody (Ab) production by Ag-specific memory B and T cells. It should be appreciated that the primary signal in models A and B is delivered by a polyclonal activator, the system involves large numbers of B cells, and is comparatively simple. In contrast, in C and D the FDC provide both the primary signal (via specific Ag) and the costimulatory signal(s). In the C and D models only

Table 2. Accessory activities of follicular dentritic cells (FDC) in vitro

Accessory activity	Experimental approach
A. FDC increase mitogen-induced B cell proliferation—FDC are active but iccosomes are not	FDC or iccosomes were cultured with B cells stimulated with LPS or anti-μ plus IL-4 and [^{3}H] thymidine uptake monitored
B. FDC increase IgG production stimulated by B cell mitogens for up to 4 weeks	FDC cultured with B cells plus LPS; Ab measured in cell culture supernatant fluid
C. Ag-bearing FDC stimulate Ag-specific B cells to proliferate (in vitro germinal centers)	Ag-bearing FDC were added to Ag-specific memory B and T cells and [^{3}H]thymidine uptake was monitored.
D. Ag-bearing FDC elicit specific IgG and IgE production; the specific IgG is produced for at least 2 months	Ag-bearing FDC were added to Ag-specific memory B and T cells and specific Ab in the supernatant fluid was monitored
E. Iccosomes bearing the appropriate Ag are potent stimulators of specific IgG production	Iccosomes added to memory B and T cells; iccosomes prepared by sonicating FDC or by shaking FDC over a filter with a larger pore size and specific Ab in the supernatant fluid monitored
F. FDC stimulate B cells to become responsive to chemotactic agents	FDC were cultured with resting B cells; the ability of B cells to chemotax after culture was assessed using Boyden chambers

Ig, immunoglobulin; Ag, antigen; LPS, lipopolysaccharide; IL, interleukin; Ab, antibody the Ag-specific B and T cells are going to be stimulated by specific antigen and repond. This is a more complicated system, but it reflects what is going on in vivo during an immune response. Parts E and F of Table 2 describe the ability of iccosomes to induce specific Ab and the ability of FDC to promote B cell chemotaxis. Each of these topics (A–F) will be covered in sequence in this review and the pertinent results discussed.

2.1 Follicular Dendritic Cells Increase Mitogen-Induced B Cell Proliferation—Iccosomes Are Not Active

Germinal centers are found in secondary lymphoid tissues and are characterized by foci of rapidly proliferating B cells surrounding Ag-bearing FDC. The germinal center reaction appears to be dependent of FDC and CD4$^+$ T cells (Tew et al. 1990, 1993). In preliminary work, it appeared that FDC were able to provide Ag-independent signaling needed for optimal proliferation of B lymphocytes (Schinzlein et al. 1984). We reasoned that lymphocytes would cluster around FDC and form a germinal center-like cluster in vitro if high-density resting B cells were activated by mitogens and cultured with FDC. This simple system facilitated study of the signals provided by FDC. In these studies lipopolysaccharide (LPS)- or anti-Ig-stimulated syngeneic or allogeneic B cells were cultured with FDC. The B cells clustered around the FDC and a two- to threefold increase in the level of proliferation was obtained (Burton et al. 1993). This costimulatory signal was dose dependent and could be ablated by removing the FDC. Interestingly, not only was proliferation increased but lymphocyte viability also appeared to be improved.

Recently we examined the effect of FDC on T cells, since these cells are also found in the germinal center microenvironment. We reasoned that although Ag retained on FDC are in large part unprocessed, FDC do possess Major histocompatibility complex (MHC) class II (Kosco et al. 1986) and might present Ag to T cells. We further reasoned that since germinal center B cells are potent antigen-presenting cells (APC) to T cells (Kosco et al. 1992), the FDC might provide accessory signaling to T cells activated via Ag presentation by germinal center B cells. When murine T cells were activated with concanavalin A and cocultured with FDC, costimulation was readily detected as evidenced by increased T cell proliferation. When FDC were removed, the increased proliferation was ablated (G.F. Burton and D. Qin, unpublished). Interestingly, FDC costimulation of T cells was not species restricted, since human FDC could costimulate mitogen-activated murine T cells and murine FDC could costimulate mitogen-activated human T cells (G.F. Burton and D. Qin, unpublished). These results may be important to human immunodeficiency virus (HIV) pathogenesis since early after infection, virus becomes localized on FDC in lymphoid follicles (Armstrong and Horne 1984; Tenner-Racz et al. 1991; Pantaleo et al. 1993). Presumably CD4$^+$ T cells coming in contact with activation signals in the germinal center would be further stimulated by FDC signaling. In addition, the FDC would provide not only a source of costimulatory signals, but also represent a source of HIV for infection of the T cells under ideal circumstances (i.e., a high degree of activation).

The costimulatory signaling mediated by FDC appears complex. Supernatant fluid derived from cultures of FDC show variable effects on B cell proliferation. If the B cells are activated using anti-Ig and interleukin (IL)-4, augmentation of proliferation has been observed using the supernatants. In contrast, if the B cells are activated by LPS, no such increase is observed (BURTON et al. 1993). The costimulatory activity for the anti-Ig and IL-4 system could be recovered from supernatant fluids harvested every other day for 2 weeks. However, if the FDC were fixed with 1.5% glutaraldehyde, no costimulatory activity could be found in the supernatant fluid and the fixed FDC per se could not provide costimulatory activity (D. QIN et al. unpublished). These data imply that costimulatory activity in the supernatant requires normal functioning of FDC and can be produced continuously for at least 2 weeks.

We know that immune complex-coated bodies (iccosomes) can be formed and released by FDC after antigen stimulation. Iccosomes have close contact with adjacent lymphocytes in vivo, and released iccosomes may be engulfed by adjacent lymphocytes (SZAKAL et al. 1988, 1989). This intimate relationship between iccosomes and lymphocytes led to the hypothesis that the iccosomes might have costimulatory activity and might be responsible for the costimulatory activity in supernatant fluids. To test this hypothesis, iccosome preparations were prepared in two ways. First, FDC were sonicated to release iccosomes and make iccosome like fragments. The iccosomes were subjected to ultracentrifugation (100 000 *g* for 2 h) and the pellet was resuspended and used as the iccosome preparation. Second, enriched FDC preparations were put onto a transwell with 3.0-μm pore size and shaken at 4°C for 3 days. This period of shaking allowed the iccosomes to separate from the FDC and to fall through pores in the transwell membrane into the lower chamber. Then the iccosome containing fluid was collected from the lower chamber of the transwell and ultracetrifuged. The pellet was resuspended and used as second source of iccosomes (J. WU et al., unpublished). We feel these iccosomes are more physiologically relevant than those created by sonication because the sonication kills the cells and releases all cellular constituents. Our two iccosome preparations were put into different B cell cultures, together with anti-μ and IL-4, and B cell proliferation was monitored. The results showed that neither sonicated FDC nor the iccosomes prepared from the transwells provided costimulatory activity. In fact, both iccosome preparations tended to inhibit B cell proliferation in the anti-μ and IL-4 culture system (D. QIN et al., unpublished).

The costimulatory activity found in the supernatant fluid after culture of FDC was also further investigated using the anti-Ig and IL-4 system. We reasoned that if the costimulatory activity in the supernatant were attributable to iccosomes, then the activity should be removed by ultracentrifugation (100 000 *g* for 2 h).

However, the activity was in the supernatant fluid and not in the pellet (containing the iccosomes) after ultracentrifugation (D. QIN et al., unpublished). This implies that costimulatory activity comes from soluble molecules and not iccosomes. However, when an active supernatant fluid was filtered through a membrane with a 10-kDa cutoff (Centricon 10), the costimulatory activity did not pass (D. QIN et al., unpublished). This indicates that the molecular mass of the costimulator found in the supernatant fluid is greater than 10 kDa, but it is clearly not the complex iccosomes.

2.2 Follicular Dendritic Cells Increase Immunoglobulin G Production Stimulated by B Cell Mitogens for up to 4 Weeks

Since FDC can increase B cell proliferation, we reasoned that FDC might also enhance antibody production. To begin testing this hypothesis, FDC were cultured with mitogen (LPS–*Salmonella typhii*)-stimulated lymphocytes. These cultures were followed for up to 2 months and IgG production was measured in the supernatant fluid taken from the cultures at various times. After 10 days of culture, wells with lymphocytes, mitogen and FDC demonstrated higher levels of IgG production (two- to ninefold increases) compared to control wells lacking the FDC (E. HAHN, unpublished). Furthermore, we were able to detect significant IgG production (>100 ng/ml) between days 17 and 24 in culture. Similarly, some cultures containing FDC continued to produce detectable antibody through a 1-month culture period (over 70 ng IgG/ml produced between days 24 and day 30; (E. HAHN, unpublished). In contrast, the amount of IgG in the control wells (lymphocytes only, lymphocytes and mitogen, FDC and mitogen) were very low to undetectable after 24 days of culture and the cells were virtually all dead (E. HAHN, unpublished). These data are consistent with the hypothesis that FDC enhance the initial burst of antibody production (measured at day 10). Furthermore, it appears that FDC may also play an important role in allowing the lymphocytes/AFC to survive and maintain antibody production in the long term.

2.3 Antigen-Bearing Follicular Dendritic Cells Stimulate Antigen-Specific B Cells to Proliferate

FDC are believed to provide important signals for the generation and development of germinal centers (TEW et al. 1993). Germinal center formation leads to the production of B memory cells (TSIAGBE et al. 1992) and in secondary responses it leads to the production of Ab-forming cells (AFC) (TEW et al. 1992b). By use of FDC in vitro, it has been possible to reconstruct germinal centers in vitro. A requirement for FDC in germinal center development comes from studies of germinal center reactions in vitro. Using the basic technique developed by our laboratory to isolate FDC (Schnizlein et al. 1985), a mixed population containing 85%–90% B cells, 3%–10% FDC, and 1%–5% T cells is obtained. This combination of cells forms small clusters of 20–50 lymphocytes per FDC within hours of culturing at 37°C. Adding [^{3}H]-thymidine at various intervals demonstrated that cells continue to proliferate for at least 5 days. Immunocytochemistry in conjunction with autoradiography revealed that the majority of cells undergoing proliferation were of the germinal center B cell phenotype (Kosco et al. 1992). The necessity of FDC and T cells was assessed by depleting these populations from the isolated cells using magnetic beads coated with FDC-specific or T cell-specific monoclonal antibodies (Kosco et al. 1992). When FDC were depleted, cell cluster formation was minimal and incorporation of [^{3}H]-thymidine was dramatically reduced. Removal of T cells similarly resulted in the formation of smaller clusters and markedly decreased proliferation (Kosco et al. 1992). The lymphocytes tend to cluster around the Ag-bearing FDC, and the B cells in the clusters proliferate (Kosco et al. 1992). Augmented B cell proliferation may also be observed with human FDC (Heinen et al. 1988; Petrasch et al. 1991, 1992). However, there is a report of human FDC inhibiting B cell proliferation (Freedman et al. 1992), suggesting that the FDC are capable of delivering multiple signals. Using an Ag-specific system, Kosco-Vilbois et al. (1993) found that murine FDC provided a primary signal to specific B lymphocytes which resulted in upregulation of their MHC class II and B7 molecules as well as Ag-presenting capability. This ability of FDC to regulate B cell proliferation and upregulate B cell molecules could be critical to the development of germinal centers in vivo.

2.4 Antigen-Bearing Follicular Dendritic Cells Elicit Specific Immunoglobulin G and E Production

Within a few days after secondary immunization, a population of AFC have been observed in developing germinal centers in vivo (Szakal et al. 1989; Kosco et al. 1989). These cells leave the germinal center and large numbers of them migrate to the bone marrow, where the vast majority of Ab in secondary responses is produced (Benner et al. 1981; Tew et al. 1992b). Furthermore, there is considerable evidence that the Ag retained on the FDC plays a major role in the long-term maintenance of serum antibody levels (Tew et al. 1990). In view of these results, it seems reasonable that the Ag on FDC should be able to induce IgG production. Experimental support for the involvement of FDC and persisting Ag in the maintenance of the immune response includes the "spontaneous induction" of IgG, which is defined as Ab production in the absence of added exogenous Ag. This spontaneous specific Ab production occurs when secondary lymphoid tissues or cells are moved from an Ab-rich environment (in vivo) into an Ab-poor environment (in vitro). In immune rabbits, such spontaneous IgG responses were elicited in cell cultures from draining lymph nodes (Greene et al. 1975; Tew et al. 1973). In rabbits, IgG spontaneous responses were obtained in vitro as late as 18 months after immunization (Stecher and Thorbecke 1967), and spleen cells of mice immunized 7 months earlier spontaneously began producing IgG upon adoptive transfer into normal irradiated recipients (Mitchison 1969). Furthermore, these spontaneous IgG responses were localized to cultures prepared from draining lymph nodes (which include FDC with persisting specific Ag). When cells from draining lymph nodes are transferred into irradiated recipients (low circulating specific Ab levels), specific IgG is induced without administration of exogenous Ag (Tew and Mandel 1978; Phipps et al. 1980). However, cells transferred from nondraining lymph nodes (contralateral to the site of immunization with no specific Ag) induced little, if any, specific IgG in the absence of added Ag (Tew and Mandel 1978).

In vitro, fragments of murine lymph nodes containing Ag-bearing FDC spontaneously produced specific IgG, confirming the adoptive transfer studies (Tew and Mandel 1978). Feedback inhibition was demonstrated in vitro through the addition of specific Ab to the culture (Tew et al. 1976). From these and other studies, it appears that Ag-bearing FDC are critical to the maintenance phase of IgG (Tew et al. 1980, 1990).

Given data implicating FDC in the long-term maintenance of serum IgG (reviewed above), we reasoned that FDC may play a role in the maintenance and regulation of IgE. To test this, mice were immunized with low doses (5-10 μg) of Ag with alum to enhance IgE responses (Vaz and Levine 1970). The following results from a number of experiments were obtained. First, like IgG (Tew and Mandel 1978), specific IgE was spontaneously induced (in the absence of exogenously added Ag) in lymph node fragment cultures derived from immunized animals (Helm et al. 1994). Secondly, like IgG (Greene et al. 1975; Tew and Mandel 1978), a spontaneous specific IgE response was elicited in vitro using cells from the draining lymph nodes (ipsilateral to the site of immunization) of immunized animals, but not from the nondraining lymph nodes (contralateral to the site of immunization) of these animals (Helm et al. 1994). Third, like IgG (Stavitsky et al. 1974), the specific IgE response elicited by Ag-bearing FDC required specific B and T memory cells (Helm et al. 1994). In these experiments, enriched preparations of Ag-specific FDC were combined with appropriate, inappropriate, or nonimmune B and T memory cells and cultured. Only cultures in which appropriately matched FDC and B and T memory cells were combined was specific IgE produced. Fourth, the specific IgE response was dependent upon FDC bearing Ag (Helm et al. 1994). FDC were deleted from preparations through use of an FDC-specific monoclonal Ab (mAb) bound to magnetic beads which, when cultured with FDC preparations, eliminated Ag-bearing FDC from the preparations. Using these preparations and appropriate B and T memory cells, the specific IgE (as well as IgG) response was ablated (Helm et al. 1994). Finally, although no real attempt was undertaken to determine the length of time a spontaneous IgE response persisted, like IgG (Tew et al. 1980), spontaneous IgE persisted for months after the final Ag challenge (up to 5 months; Helm et al. 1994). Interestingly, like the mitogen system discussed in Sect. 2.2 above, the B and T memory cells cluster around the FDC and Ab production can be detected for at least 2 months (D. Qin, unpublished). In short, the results discussed here support the concept that the FDC are important accessory cells involved in maintaining serum IgG and IgE.

2.5 Iccosomes Bearing the Appropriate Antigen Are Potent Stimulators of Specific Immunoglobulin G Production

We know that iccosomes can be formed and released by FDC after Ag stimulation. Furthermore, iccosomes have close contact with adjacent lymphocytes in vivo, and released iccosomes may be engulfed by adjacent lymphocytes (SZAKAL et al. 1989). We reasoned that the iccosomes might be involved in the elicitation of specific Ab production in vivo. To test this hypothesis, iccosomes were prepared in two ways, as described in Sect. 2.1 above. Our two iccosome preparations were cultured with memory T and B cells from the spleens of ovalbumin (OVA)-immunized BALB/c mice. The culture supernatants were monitored at 6 and 12 days for anti-OVA IgG production using a modified enzyme-linked immunosorbent assay (ELISA). When memory T and B cells were cultured alone, only a low background level of anti-OVA IgG was recovered in their supernatant. In contrast, when either iccosome preparation was added to the culture system anti-OVA IgG production was markedly higher (J. WU et al., unpublished). Interestingly, iccosomal Ag was much more potent that the same amount of free Ag (OVA) or preformed immune complex (OVA–anti-OVA). Other controls included iccosomes bearing irrelevant Ag which failed to induce IgG–anti-OVA production and placing FDC on a transwell with a 0.1-μm pore size which would not allow iccosomes to come through (J. WU et al., unpublished). These results provide support for the hypothesis that iccosomes may play an important role in stimulating B cells to produce specific Ab. However, it appears that their function is more than simply providing Ag, since free Ag or Ag–Ab immune complexes failed to produce comparable responses.

2.6 Follicular Dendritic Cells Stimulate B Cells to Become Responsive to Chemotactic Agents

The bulk of Ab production in mammals occurs in the bone marrow, yet Ag stimulation of B cells occurs in secondary lymphoid tissues (BENNER et al. 1981; TEW et al. 1992b). We believe that during the early phase of the germinal center reaction (days 1-5 after administering Ag), germinal center B cells receive the signaling needed to develop into AFC and then migrate from the germinal center through the lymph and blood to the bone marrow, where they terminally differentiate and secrete specific Ab (BENNER et al. 1981; TEW et al. 1992b). We further reasoned that if the germinal center-derived AFC were chemotactically active, it would aid in migration from the blood vasculature in the bone marrow into the bone marrow proper. As These AFC travel through the vasculature, they may be able to recognize adhesion molecules/molecular addressins present on endothelium and adhere to these sites (TAVASSOLI and HARDY 1990; KUPP et al.

1991). Once bound to the addressins, they could respond to chemotactic signals present in the microenvironment and travel up this gradient into the bone marrow to perform their effector function (i.e., Ab secretion). We have made a number of observations which support this model:

1. During the early phase of the germinal center reaction, AFC with the characteristics of germinal center B cells (e.g., blasting phenotype, PNA^{Hi}, $B220^{+}$) can be observed in apparent migration from germinal centers.
2. Germinal center B cells capable of spontaneous Ab secretion can be found in the peripheral blood and lymph 3 days after Ag challenge but not at later times. Furthermore, these cells secrete Ab specific for Ag retained on FDC in draining lymph nodes (DiLosa et al. 1991; Tew et al. 1992b).
3. Adoptive transfer of germinal center B cells isolated 3 days after Ag challenge (but not at 10 days) results in homing to the bone marrow (Tew et al. 1992b).
4. Germinal center B cells isolated from draining lymph nodes during the early phase of the germinal center reaction are chemotactically active and will chemotax toward factors produced in bone marrow cultures (Kupp et al. 1991; Tew et al. 1992b).

Since the signaling required for generation of AFC appeared to be occurring in germinal centers, we reasoned that FDC might provide signals to help germinal center B cells to become chemotactically active. To test this, FDC were cocultured with non-migrating, high-density B cells in the presence or absence of activation signals (anti-Ig and IL-4). Following culture, the B cells were examined for their ability to respond to chemotactic signals. After 3 days of coculture with FDC, B cells demonstrated both chemotaxis and chemokinesis. This response did not require preactivation by anti-Ig and/or IL-4, nor could it be augmented by any combination of these agents. Induction of B cell chemotaxis occurred within 6 h of culture with FDC and reached a peak after 2 days. Removal of FDC ablated the inductive event, confirming that FDC were responsible. Furthermore, the FDC could not induce B cell chemotactic ability through a transwell, indicating that FDC–B cell contact is probably required. FDC from athymic (nude) or euthymic mice induced equivalent responses, indicating that functional T cells were not required (G. Burton, unpublished). These findings support the concept that FDC-mediated signals to B cells not only activate these cells and increase their proliferative potential, but also help facilitate dissemination of these cells to sites where full effector function can be developed.

References

Armstrong JA, Horne R (1984) Follicular dendritic cells and virus-like parcticles in AIDS-related lymphadenopathy. Lancet 2:370
Benner R, Hijmans W, Haaijman JJ (1981) The bone marrow: the major source of serum immunoglobulins, but still a neglected site of antibody formation. Clin Exp Immunol 46: 1–8
Berek C, Ziegner M (1993) The maturation of the immune response. Immunol Today 14: 400–404

Burton GF, Conrad DH, Szakal AK, Tew JG (1993) Follicular dendritic cells (FDC) and B cell co-stimulation. J. Immunol 150: 31–38

Burton GF, Kapasi ZF, Szakal AK, Tew JG (1994) The generation and maintenance of antibody and B cell memory. In: Ada GL (ed) Vaccine strategies to control infections. Landes, Austin, p 35

DiLosa RM, Maeda K, Masuda A, Szakal AK, Tew JG (1991) Germinal center B cells and antibody production in the bone marrow. J Immunol 1460: 4071–4077

Freedman AS, Munro JM, Rhynhart K, Schow P, Daley J, Lee N, Svahn J, Eliseo L, Nadler LM (1992) Follicular dendritic cells inhibit human B-lymphocyte proliferation. Blood 80: 1284–1288

Greene EJ, Tew JG, Stavitsky AB (1975) The differential localization of the in vitro spontaneous antibody and proliferative responses in lymphoid organs proximal and distal to the site of primary immunization. Cell Immunol 18: 476–483

Heinen E, Cormann N, Lesage F, Kinet-Denoel C, Tsunoda R, Simar LJ (1988) Follicular dendritic cells act as accessory cells. In: Schook LB, Tew JG (eds) Antigen presenting cells diversity, differentiation, and regulation. Liss, New York, p 69

Helm SLT, Burton GF, Szakal AK, Tew JG (1994) Follicular dendritic cells and the maintenance of IgE. J Immunol (submitted)

Humphrey JH, Grennan D (1982) Isolation and properties of spleen follicular dendritic cells. In: Nieuwenhuis P van der Broek A A, Hanna MG Jr (eds) In vivo immunology. Plenum, New York, p 823

Kapasi ZF, Burton GF, Schultz LD, Tew JG, Szakal AK (1993) Induction of functional follicular dendritic cell development in severe combined immunodeficiency mice. J Immunol 150: 2648–2658

Kosco MH, Tew JG, Szakal AK (1986) Antigenic phenotyping of isolated and in situ rodent follicular dendritic cells (FDC) with emphasis on the ultrastructural demonstration of Ia antigens. Anat Rec 215: 201–213

Kosco MH, Burton GF, Kapasi ZF, Szakal AK, Tew JG (1989) Antibody-forming cell induction during an early phase of germinal centre development and its delay with ageing. Immunology 68: 312–318

Kosco MH, Pflugfelder E, Gray D (1992) Follicular dendritic cell-dependent adhesion and proliferation of B cells in vitro. J Immunol 148: 2331–2339

Kosco-Vilbois MH, Gray D, Scheidegger D, Julius M (1993) Follicular dendritic cells help resting B cells to become effective antigen presenting cells: induction of B7/BB1 and upregulation of MHC class II molecules. J Exp Med 178:2055–2066

Kupp LI, Kosco MH, Schenkein HA, Tew JG (1991) Chemotaxis of germinal center B cells in response to C5a. Eur J Immunol 21: 2697–2701

Lilet-Leclercq C, Radoux D, Heinen E, Kiner-Denoel C, Defraigne JO, Houben-Defresne MP, Simar LJ (1984) Isolation of follicular dendritic cells from human tonsils and adenoids. I. procedure and morphological characterization. J Immunol Methods 66: 235

MacLennan ICM, Gray D (1982) Antigen-driven selection of virgin and memory B cells. Immunol Rev 91: 61

Masuda A, Burton GF, Fuchs BA, Bhogal BS, Rupper R, Szakal AK, Tew JG (1994) Follicular dendritic cell function and murine AIDS. Immunology 81: 41–46

Mitchison NA (1969) Unmasking of cell-associated foreign antigens during incubation of lymphoid cells. Isr J Med Sci 5: 230–234

Pantaleo G, Graziosi C, Demarest JF, Butini L, Montroni M, Fox CH, Orenstein JM, Kotler DP, Fauci AS (1993) HIV infection is active and progressive in lymphoid tissue during the clinically latent stage of disease. Nature 362: 355–358

Petrasch S, Kosco MH, Perez-Alvarez CJ, Schmitz J, Brittinger G (1991) Proliferation of germinal center B lymphocytes in vitro by direct membrane contact with follicular dendritic cells. Immunobiology 183: 451–462

Petrasch S, Kosco MH, Perez-Alvarez C, Schmitz J, Brittinger G (1992) Proliferation of non-Hodgkin-lymphoma lymphocytes in vitro is dependent upon follicular dendritic cell interactions. Br J Haematol 80: 21–26

Phipps RP, Tew JG, Miller GA (1980) A murine model for analysis of spontaneous induction and feedback regulation of specific antibody synthesis. Immunol Commun 9: 55–70

Schnizlein CT, Szakal AK, Tew JG (1984) Follicular dendritic cells in the regulation and maintenance of immune responses. Immunobiology 168: 391–402

Schnizlein CT, Kosco MH, Szakal AK, Tew JG (1985) Follicular dendritic cells in suspension: identification, enrichment, and initial characterization indicating immune complex trapping and lack of adherence and phagocytic activity. J Immunol 134: 1360–1368

Stavitsky AB, Tew JG, Harold WW (1974) Thymus dependence of the spontaneous induction of the in vitro anamnestic antibody responses to human serum albumin and keyhole limpet hemocyanin. J Immunol 113: 2045–2047

Stecher VJ, Thorbecke GJ (1967) Gammaglobulin and antibody formation in vitro. VI. Effect of X-irradiation on the secondary antibody response in vitro. J Exp Med 125: 33–44
Szakal AK, Tew JG (1992) Follicular dendritic cells: B cell proliferation and maturation. Cancer Res 52: 5554s–5556s
Szakal AK, Kosco MH, Tew JG (1988) A novel in vivo follicular dendritic cell-dependent iccosome-mediated mechanism for delivery of antigen to antigen-processing cells. J Immunol 140: 341–353
Szakal AK, Kosco MH, Tew JG (1989) Microanatomy of lymphoid tissue during the induction and maintenance of humoral immune responses: structure function relationships. Annu Rev Immunol 7: 91–109
Szakal AK, Burton GF, Smith JP, Tew JG (1991a) Antigen processing and presentation in vivo. In: Spriggs DR, Koff WC (eds) Topics in vaccine adjuvant research. CRC Press, Boca Raton, p 11
Szakal AK, Kosco MH, Smith JP, Tew JG (1991b) Kinetics of the antigen transport-FDC-iccosome-B cell axis. In: Racz P, Tenner-Racz K, Dijkstra CD, Gluckman JC (eds) Accessory cells in HIV and other retroviral infections. Karger, Basel, p 29
Szakal AK, Kapasi ZF, Masuda A, Tew JG (1992) Follicular dendritic cells in the alternative antigen transport pathway: microenvironment, cellular events, age and retrovirus related alterations. Semn Immunol 4: 257–265
Tavassoli M, Hardy CL (1990) Molecular basis of homing on intravenously transplanted stem cells to the marrow. Blood 76: 1059
Tenner-Racz K, Racz P, Schmidt H, Taveres LM, de Noronha F, Stahl-Henning C, Hunsmann G (1991) Virus trapping by follicular dendritic cells in retrovirus infections inducing follicular hyperlplasia of lymph nodes. In: Racz P, Dijkstra CD, Gluckman JC (eds) Accessory cells in HIV and other retroviral infections. Karger, Basel, p 83
Tew JG, Mandel TE (1978) The maintenance and regulation of serum antibody levels: evidence indicating a role for antigen retained in lymphoid follicles. J Immunol 120: 1063–1069
Tew JG, Self CH, Harold WW, Stavitsky AB (1973) The spontaneous induction of anamnestic antibody responses in lymph node cell cultures many months after primary immunization. J Immunol 111: 416–423
Tew JG, Greene EJ, Makoski MH (1976) In vitro evidence indicating a role for the Fc region of IgG in the mechanism for the long-term maintenance and regulation of antibody levels in vivo. Cell Immunol 26: 141–152
Tew JG, Phipps RP, Mandel TE (1980) The maintenance and regulation of the humoral immune response: persisting antigen and the role of follicular antigen-binding dendritic cells as accessory cells. Immunol Rev 53: 175–201
Tew JG, Kosco MH, Szakal AK (1989) The alternative antigen pathway. Immunol Today 10: 229–231
Tew JG, Kosco MH, Burton GF, Szakal AK (1990) Follicular dendritic cells as accessory cells. Immunol Rev 117: 185–211
Tew JG, Maeda K, Imai Y, Burton GF, Szakal AK (1991) What is a follicular dendritic cell? In: Imai Y, Tew JG, Hoefsmit ECM (eds) Dendritic cells in lymphoid tissue. Elsevier Science, Amsterdam, p 19
Tew JG, Burton GF, Masuda A, Kapasi ZF, Szakal AK (1992a) Dendritic cells as accessory cells. In: Fornusek L, Vetvicka V (eds) Immune system accessory cells. CRC Press, Boca Raton, p 131
Tew JG, DiLosa RM, Burton GF, Kosco MH, Kupp LI, Masuda A, Szakal AK (1992b) Germinal centers and antibody production in bone marrow. Immunol Rev 126: 1–14
Tew JG, Burton GF, Kupp LI, Szakal A (1993) Follicular dendritic cells in germinal center reactions. In: Kamperdijk EWA, Nieuwenhuis P, Hoefsmit ECM (eds) Dendritic cells in fundamental and clinical Immunology. Plenum, New York, p 461 (Advances in Experimental Medicine and Biology, vol 329)
Tew JG, Burton GF, Szakal AK (1994) Follicular dendritic cells in antibody responses. In: Nossal GJV (ed) Handbook of experimental immunology, the lymphoid system antibody responses and affinity maturation. Rockwell scientific, Oxford (in press)
Tsiagbe VK, Linton PJ, Thorbecke GJ (1992) The path of memory B-cell development. Immunol Rev 126: 113–141
Vaz NM, Levine BB (1970) Immune responses of inbred mice to repeated low doses of antigen. Relationship to histocompatibility (H-2) type. Science 168: 852

Regulation of Human B Cell Activation by Follicular Dendritic Cell and T Cell Signals

G. Grouard, O. de Bouteiller, C. Barthelemy, S. Lebecque, J. Banchereau, and Y.-J. Liu

1 Introduction

Follicular dendritic cells (FDC) represent a unique cell type (Nossal et al. 1968; Szakal and Hanna 1968) within the B cell follicles of the secondary lymphoid tissues. During T cell-dependent humoral immune responses, B cells undergo proliferation, somatic mutation, isotype switching, positive selection, and differentiation into memory B cells and plasma cells within the FDC networks (Berek 1992; Kosco and Gray 1992; Kroese et al. 1987; Liu et al. 1992; MacLennan 1994; Nossal 1992). The close physical contact between FDC and B cells that occurs when B cells undergo dramatic phenotypic and genetic changes suggests that FDC may be directly involved in all these events. Based on the histological observations depicted in Fig. 1, follicular B cell response can be divided into three distinct stages, in which different B cell events occur:

1. Exponential growth of germinal center precursor cells within the primary follicles. Germinal center reaction is initiated by the rapid proliferation of about

---

Schering-Plough, Laboratory for Immunological Research, 27 chemin des Peupliers, B.P. 11, 69571 Dardilly Cedex, France

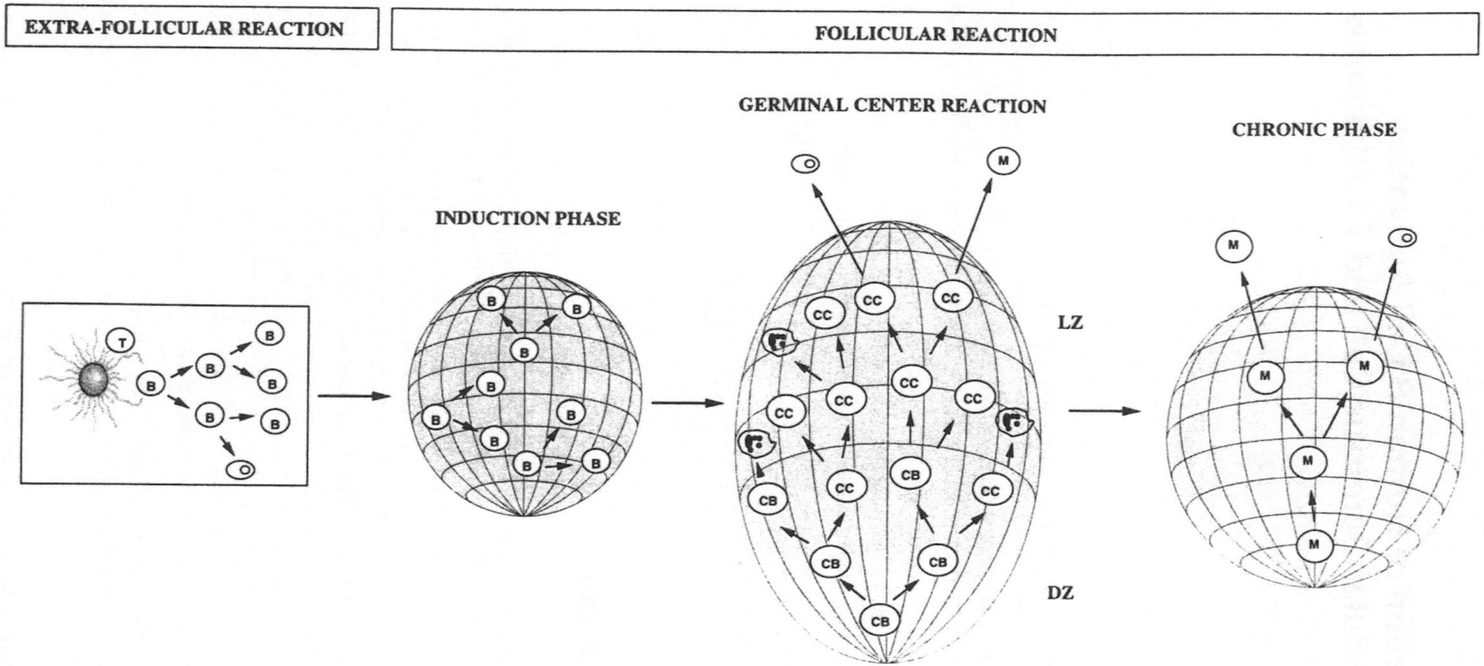

EXTRA-FOLLICULAR REACTION
FOLLICULAR REACTION
INDUCTION PHASE
GERMINAL CENTER REACTION
CHRONIC PHASE
T
B
CC
CB
M
LZ
DZ

three precursor cells (Kroese et al. 1987; Liu et al. 1991; Jacob and Kelsoe 1992), giving rise to a germinal center containing about 1.4 × 10^4 cells after 4 days of immunization (Liu et al. 1991).

2. A fully developed germinal center has a dark zone and a light zone. Centroblasts in the dark zone proliferate and mutate (Berek et al. 1991; Jacob et al. 1991; Küppers et al. 1993; McHeyzer-Williams et al. 1993). They give rise to non-proliferating centrocytes in the light zone. The high-affinity centrocytes are positively selected (Weiss et al. 1992) by antigens deposited on FDC and differentiates to either memory B cells (McHeyzer-Williams et al. 1991; Schittek and Rajewsky 1992; Weiss and Rajewsky 1990) or plasma cells (Kosco et al. 1989). The low-affinity centrocytes which have failed to be selected die by apoptosis (Liu et al. 1989).
3. After 3–4 weeks of immunization, fully developed germinal centers are replaced by small follicular foci which consist of a few antigen-specific B blasts associated with immune complex networks (Liu et al. 1991). These foci may represent the chronic stimulation of memory B cells, which are fundamental for the maintainance of serum immunoglobulin (Ig) level (Tew et al. 1980) and long-lived memory B cell clones (Gray and Skarvall 1988).

Evidence also suggests that FDC increase in cell number (Szakal et al. 1985) and change their phenotype during germinal center reaction. When a follicular reaction develops into a full germinal center, two subsets of FDC can be identified: $CD23^+$ FDC in the apical light zone of germinal center and $CD23^-$ FDC in the basal light zone and apical dark zone of the germinal center (Hardie et al. 1993). Later, when germinal center reaction decreases in size and is replaced by chronic foci of a few proliferating B blasts, fewer FDC can be identified. In addition, antigen-specific $CD4^+$ $CD45RO^+$ CD40 $ligand^+$ T cells (Fuller et al. 1993; Lederman et al. 1992) and tingible body macrophages are also recruited into the germinal centers.

Thus, the germinal center reaction represents a dynamic multicellular event. To understand the molecular mechanisms of germinal center reaction, we have followed two major approaches:

Fig. 1. Three distinct stages of follicular B cell activation (Liu et al. 1991). *Induction phase:* germinal center precursor cells were generated in the T cell (*I*) and interdigitating cell zones during the activation of immunoglobulin $(Ig)D^+$ CD 38^- naive B cells (*B*). About three precursor cells colonize a follicle and undergo exponential growth within the follicular dendritic cell (FDC) network for about 3 days. These proliferating cells might have a IgD^+ $CD38^+$ phenotype in human tonsils. Selection and somatic mutation start at this stage (Y.-J. Liu et al., manuscript in preparation). *Germinal center stage:* fully developed germinal centers can be recognized at about 4 days after immunization in carrier-primed rats. Centroblasts (*CB*) in the dark zone of the germinal center continue to proliferate, mutate, and undergo selection. They give rise to nonproliferating centrocytes (*CC*), forming the light zone of the germinal centers. They undergo affinity selection based on their binding to antigens retained on FDC. Only high-affinity centrocytes survive, preferentially expand, and finally differentiated into memory B cells or plasma cells. Somatic mutation, isotype switching, and selection may operate simultaneously and continuously before high-affinity centrocytes leave the germinal centers. *Chronic phase:* germinal centers are replaced by chronic foci of antigen-specific B cell proliferation within the follicles. This may continue to give rise to memory B cells (*M*) and plasma cells (*LZ*, light zone; *DZ*, dark zone)

1. Isolation of naive B cells, germinal center precursor cells, germinal center B cells, and memory B cells. These different B cell subsets may respond differently to signals from FDC.
2. Study of the effect of T cell-derived soluble cytokines and membrane-bound molecules on FDC-dependent B cell activation.

This paper represents a brief review on our progress.

2 Anti-Immunoglobulin D and Anti-CD38 Double Staining Identifies Four Major Tonsillar B Cell Subsets

On tonsil sections, anti-IgD stains naive B cells in the follicular mantle, while anti-CD38 stains germinal center B cells and plasma cells (Fig. 1). When total tonsillar B cells were analyzed by anti-IgD–phycoerythrin(PE) and anti-CD38–fluorescein isothiocyanate (FITC) double staining (Fig. 2), IgD^+CD38^- naive follicular mantle B cells and IgD^- $CD38^+$ germinal center B cells were clearly identified. In addition, an IgD^- $CD38^-$ double negative subset and an IgD^+ $CD38^+$ double positive subset was identified. These two additional subsets of B cells may represent the memory B cell subset and germinal center precursor cell subset, respectively (Liu et al., in press).

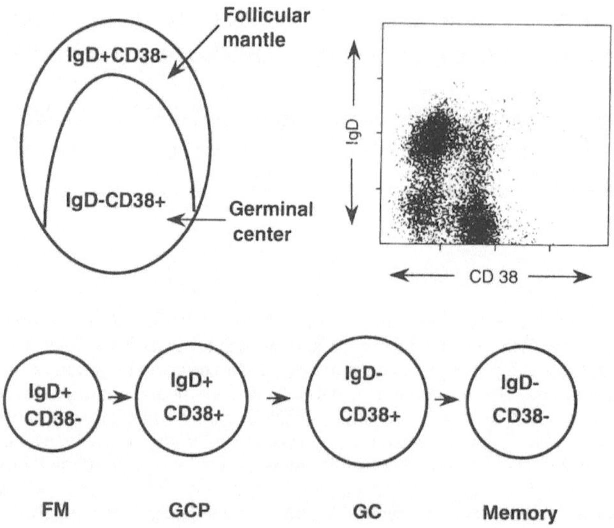

Fig. 2. Identification of immunoglobulin (*Ig*) D^+ $CD38^-$ naive B cells, IgD^+ $CD38^+$ germinal center (*GC*) precursor cells, and IgD^- $CD38^+$ germinal center B cells, and IgD^- $CD38^-$ memory B cells. *FM*, follicular mantle; *GCP*, germinal center precursor

3 Immunoglobulin D$^-$ CD38$^-$ B Cells Are Memory B Cells

The IgD$^-$ CD38$^-$ B cell subset can be isolated from total tonsil B cells by negative magnetic beads depletion of IgD$^+$ B cells and CD38$^+$ B cells. Detailed phenotypic fluorescence-activated cell sorter (FACS) analysis shows that IgD$^-$ CD38$^-$ double negative B cells do not express the follicular mantle B cell markers IgD and CD23 or germinal center B cell markers CD38, CD10, and CD77 (Table 1). They express high levels of CD39, CD44, and surface IgG (Table 1). This phenotypic characteristic suggests that IgD$^-$ CD38$^-$ B cells might be memory B cells rather than germinal center precursor cells, since most of these cells have undergone isotype switching. Accordingly, these IgD$^-$ CD38$^-$ cells have undergone somatic hypermutation in their IgV genes of both IgM and IgG transcripts, demonstrating their germinal center origin (Pascual et al. 1994); they are resting cells which express Bcl-2 protein and are resistant to apoptosis in vitro, suggesting that they have been selected within germinal centers; in addition, they have the ability to produce specific antibody to tetanus toxoid after polyclonal stimulation. Finally, in situ double staining with anti-IgD$^+$ anti-CD38 (blue color) and anti-CD20 (red color) showed that IgD$^-$ CD38$^-$ CD20$^+$ memory B cells were located mainly within the intraepithelial areas of mucosal surface (Liu et al., submitted).

4 Immunoglobulin D$^+$ CD38$^+$ Double Positive B Cells Are Germinal Center Precursor Cells

Three-color FACS analysis allows us to characterize the detailed phenotype of the IgD$^+$ CD38$^+$ double-positive subset of B cells. They have all the features of germinal center centroblasts, except that they express surface IgD (Fig. 3): (a) they express germinal center markers CD10, CD38, CD71, and CD77; (b) they are

Table 1. Phenotypic and genetic analysis of the four tonsillar B cell subsets

	FM	GCP	GC	Memory
CD23	30%+	–	–	–
CD39	+	?	–	+
CD44	+	Low	Low	+
CD10	–	+	+	–
CD71	–	+	+	–
CD77	–	50%+	40%+	–
IgM	+	+	10%+	5%+
IgG	–	–	+	+
Bcl-2	+	–	–	+
fas	–	+	+	–
Ki67	–	+	+	–
Somatic mutation	–	70%+	+	+

Ig, immunoglobulin; *FM*, follicular mantle; *GCP*, germinal center precursor; *GC*, germinal center.

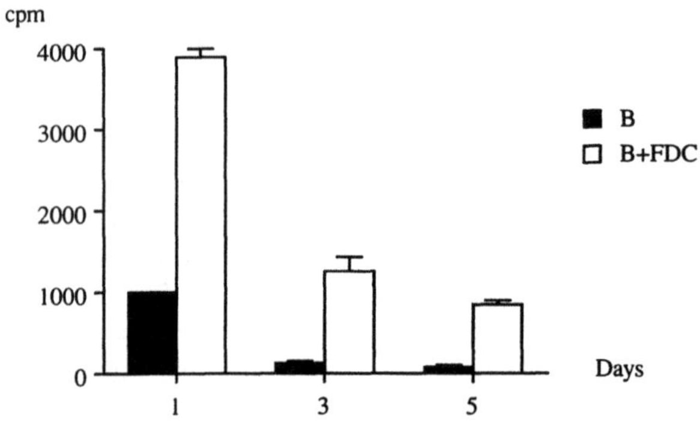

Fig 3. Kinetics of B cell proliferation cultured with and without follicular dendritic cells (*FDC*). A total of 10^5 autologous B cells were cultured with 2000 FDC clusters (2000 rad irradiation) for 1, 3, and 5 days; 1µ Ci[^{3}H]-thymidine was added 8 h before cell harvesting. The thymidine incorporation was expressed as counts per minute (cpm) ± S.D. of triplicated wells. The control thymidine incorporation by irradiated FDC was 298±12

cycling cells as determined by intracellular Ki-67 staining and Hoechst 33342 staining; and (c) they do not express survival gene *bcl*-2, but express death gene *fas*. During in vitro culture, IgD$^+$ CD38$^+$ double-positive cells undergo rapid apoptosis, as do IgD$^-$ CD38$^+$ germinal center B cells. Sequence analysis of their IgV genes shows that 50% of both IgM and IgD transcripts were mutated. By double anti-IgD and anti-CD38 staining on tonsil section, IgD$^+$ CD38$^+$ B cells were located mainly in the dark zone of germinal centers. The above experimental data indicate that IgD$^+$ CD38$^+$ double-positive B cell subsets may represent germinal center precursor cells which initiate germinal center reaction within the FDC network (Y.-J. Liu and S. Lebecque, manuscript in preparation).

5 Human Follicular Dendritic Cells Can Enhance B Cell Proliferation

FDC have been shown to have a stimulatory effect on B cells in both human and mouse systems by many groups (Cormann et al. 1986; Kosco et al. 1992; Burton et al. 1993; Petrasch et al. 1991). In contrast, the experiments by Freedman et al. (1992) demonstrated that FDC inhibit human B cell proliferation. Our recent data in human system shows that FDC clusters from human tonsils promote moderate and short-term autologous B cell proliferation (Fig. 3), which can be blocked completely by removing FDC by FDC-specific monoclonal antibody 7D6 generated in the laboratory (Fig. 4c; G. Grouard et al., manuscript in preparation). The experiments by Petrasch et al. (1991) showed that the proliferating B cells in FDC-dependent culture were always in close contact with FDC. Whether the stimulatory signal from FDC represent soluble factors or membrane molecules is presently unknown. The adhesion of B cells onto FDC was shown to be mediated a

b

c

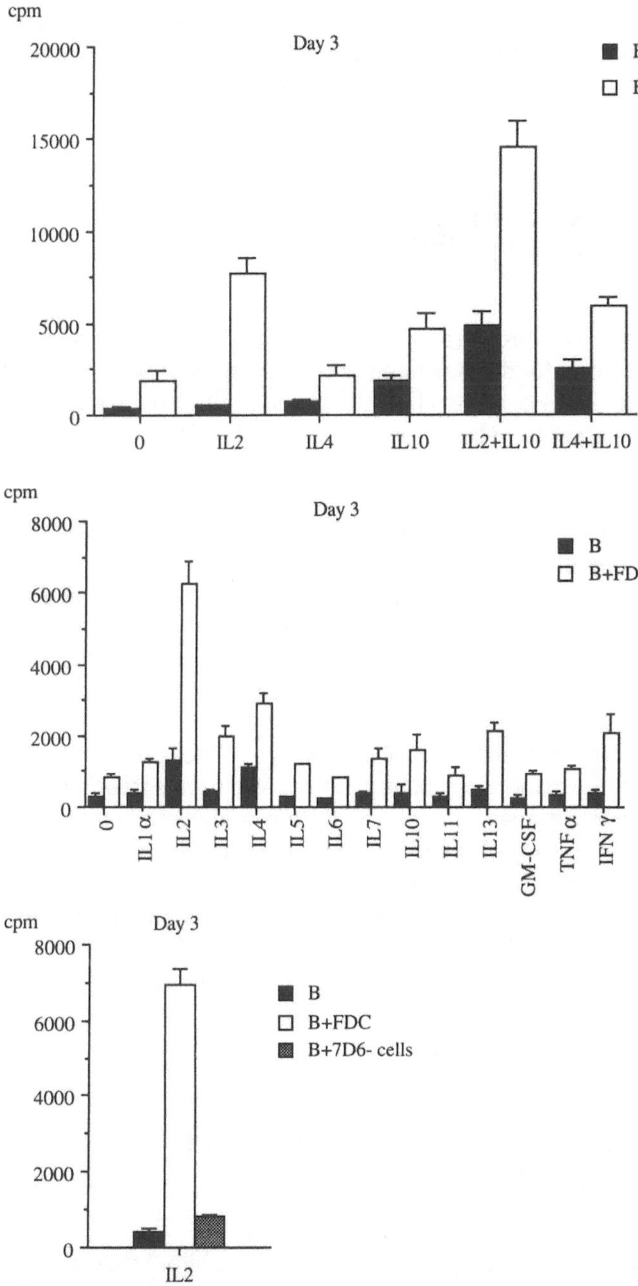

Fig. 4. a–c The effect of soluble cytokines on follicular dendritic cell (FDC)-dependent B cell proliferation. The control thymidine uptake by irradiated FDC with interleukin (*IL*)-2 was 247±52. *GM-CSF*, granulocyte-macrophage colony-stimulating factor; *TNF*, tumor necrosis factor; *IFN*, interferon. **b** IL-2 (10 units/ml) and IL-10 (100ng/ml) have an additive effect on FDC-dependent B cell proliferation. The control thymidine uptake by irradiated FDC with IL-2 and IL-10 is 224±40. **c** FDC-dependent B cell proliferation in the presence of IL-2 was completely abolished by depleting FDC with monoclonal antibody (mAb) 7D6 and anti-mouse immunoglobulin (Ig)-magnetic beads

by adhesion molecules very late antigen (VLA)-4, vascular cell adhesion molecule (VCAM)-1, lymphocyte function-associated antigen (LFA)-1, and intercellular adhesion molecule (ICAM)-1 (Freedman et al. 1990; Koopman et al. 1991; Kosco 1991).

6 Interleukin-2 Enhances Follicular Dendritic Cell-Dependent B Cell Proliferation

Whether FDC provide unique proliferation signals for germinal center B cells or only provide survival signals for the proliferating cells is an important question to address. A recent study by Lindhout et al. (1993) provides direct evidence in vitro that FDC rescue germinal center B cells from apoptosis. The FDC-derived survival factors may include antigens (Liu et al. 1989), complement fragments (Bonnefoy et al. 1993), and adhesion molecule ICAM-1 (Koopman et al. 1994). To achieve significant FDC-dependent B cell proliferation, Burton et al. (1993) added T cell cytokine interleukin (IL)-4 plus anti-IgM or lipopolysaccharide (LPS) into the cell culture, and Cormann et al. added anti-IgD and phorbol myristate acetate (PMA) into the culture. The experiment by Kosco et al. (1992) showed that the spontaneous B cell proliferation within the FDC clusters can be totally blocked by depleting T cells. Taken together, these experiments suggest that T cells may play a key role in FDC-dependent B cell proliferation.

Accordingly, we have demonstrated the presence of T cells within the FDC clusters (G. Grouard et al., manuscript in preparation). We have screened a wide range of soluble cytokines on their effect on FDC-dependent B cell proliferation. Figure 4a shows that IL-2 is the only cytokine which gives consistently moderate enhancement of FDC-dependent B cell proliferation. IL-2 and IL-10 have some additive effects (Fig. 4b; G. Grouard et al., manuscript in preparation). However, the limited FDC-dependent B cell proliferation cultured with IL-2 or IL-2 plus IL-10 suggests that additional signals are required to mimic the intense germinal B cell proliferation.

7 CD40–CD40 Ligand Interaction Is Important for Germinal Center Reaction

Anti-CD40 antibody was initially recognized by its strong costimulatory effect for B cell proliferation with anti-IgM, phorbol ester (Clark and Ledbetter 1986; Ledbetter et al. 1987; Vallé et al. 1989), or IL-4 (Gordon et al. 1988). Interestingly, anti-CD40 antibody G28.5 was found to deliver the strongest signal to prevent germinal center B cells from apoptosis (Liu et al. 1989). In 1991 Banchereau et al. found that anti-CD40 antibody presented by mouse L cells transfected with human FcγRII/CD32 gene plus IL-4 could permit human B cell growth for 5 weeks with 1000-fold expansion in cell number. It was then suggested that this in vitro culture system might partially mimic germinal center reaction (BANCHEREAU and ROUSSET 1991). It was also suggested that the ligand for CD40 antigen might be within the germinal centers (CLARK 1990). Molecular cloning established that CD40 ligand is a member of the tumor necrosis factor (TNF) cytokine superfamily and is expressed on early-activated T cells.

The direct evidence that CD40–CD40 ligand interaction is important for germinal center reaction has been provided very rapidly by several groups:

1. In human tonsil sections, CD40 ligand is mainly expressed on germinal center T cells (LEDERMAN et al. 1992), thus providing direct in vivo evidence that CD40–CD40 ligand interaction does occur within the human germinal centers. CD40 ligand has also been detected in T cell zones when humoral immune responses are initiated (VAN DEN EERTWEGH et al. 1993)
2. Hyper-IgM patients display mutations in CD40 ligand genes; their T cells cannot express functional CD40 ligand. The secondary lymphoid tissues of these patients lack germinal centers (ALLEN et al. 1993; ARUFFO et al. 1993; DISANTO et al. 1993; FULEIHAN et al. 1993; KORTHÄUER et al. 1993).
3. Hyper-IgM syndrome has been reproduced in mouse models by disrupting CD40 genes (KAWABE et al. 1994) or by injecting anti-CD40 ligand antibodies to block CD40–CD40 ligand interaction in vivo during T cell-dependent immune responses (FOY et al. 1994). Germinal centers cannot be identified in these mice.

8 CD40 Ligation Is Essential for Maximal Follicular Dendritic Cell-Dependent B Cell Proliferation In Vitro

We have studied the direct effect of CD40 ligation on FDC-dependent B cell proliferation. Figure 5 shows that anti-CD40 antibody G28.5, in combination with IL-2, IL-3, IL-4, IL-10, and IL-13, significantly potentiates FDC-dependent B cell proliferation. Maximal proliferation was obtained with the combination of anti-CD40, IL-2, and IL-10 or anti-CD40, IL-4, and IL-10. Proliferation peaks at day 3 of culture (GROUARD et al., manuscript in preparation).

9 Conclusion and Future Prospects

Anti-IgD and anti-CD38 double staining permits isolation of IgD^+ $CD38^-$ naive B cells, IgD^+ $CD38^+$ germinal center precursor cells, IgD^- $CD38^+$ germinal center precursor cells in the dark zone of the germinal center are found to undergo

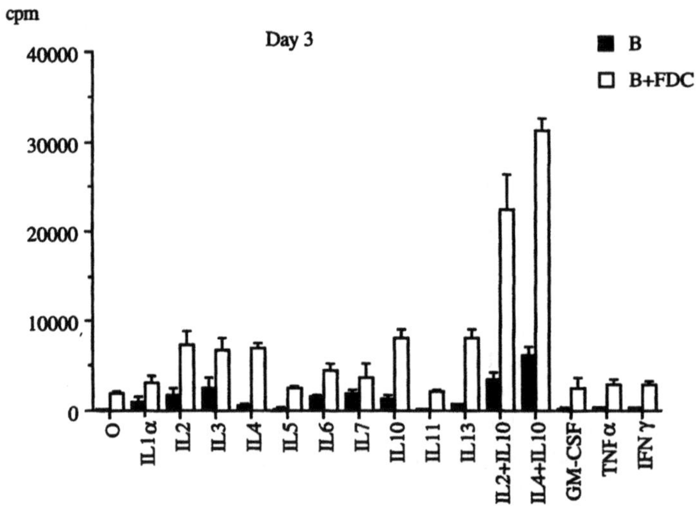

Fig. 5. Maximal follicular dendritic cell (FDC)-dependent B cell proliferation was observed when anti-CD40 plus interleukin (*IL*)-2 plus IL-10 and anti-CD40 plus IL-4 (50 units/ml) plus IL-10 were added to the culture. A total of 0.5×10^5 B cells were cultured with 2000 FDC; 1 μg anti-CD40 antibody G28.5/ ml (a kind gift from Dr. E. Clark) was used in the culture. The thymidine uptake by irradiated FDC cultured with anti-CD40 plus IL-4 plus IL-10 was 703±376. *GM-CSF*, granulocyte-macrophage colony-stimulating factor; *TNF*, tumor necrosis factor; *IFN*, interferon

proliferation, apoptosis, and somatic mutation. Thus the study of the interaction of this B cell population with follicular dendritic cells and T cells may provide important information regarding mechanisms of the induction of germinal center reaction. Large numbers of $IgD^- CD38^-$ memory B cells are also isolated. We will therefore compare the reactivities of $IgD^+ CD38^-$ naive B cells and $IgD^- CD38^-$ memory B cells with FDC in vitro.

We have confirmed earlier reports that FDC could stimulate B cell proliferation, but in a limited fashion. Further ligation of CD40 antigen in combination with IL-2 and IL-10 or IL-4 and IL-10 is essential for maximal FDC-dependent B cell proliferation in vitro. Together with the data derived from hyper-IgM patients and CD40-knockout mice, this strongly suggests that cognate B cell–FDC–T cell interaction within the follicles is requied for germinal center reaction. Whether FDC can provide unique proliferation signals or only provide survival signals for the proliferating B cells is an open question that we are currently addressing.

Acknowledgment. We would like to thank Mrs. N. Courbière and Muriel Vatan for their excellent editorial help. Gèraldine Grouard is supported by the Foundation Mérieux. We wish to thank our collaborators J. Donald Capra, Virginia Pascual, and Antony Magalski for sequencing Ig genes, Dr. Edward Clark for G28.5 Ab.

References

Allen RC, Armitage RJ, Conley ME, Rosenblatt H, Jenkins NA, Copeland NG, Bedell MA, Edelhoff S, Disteche CM, Simoneaux DK, Fanslow WC, Belmont J, Spriggs MK (1993) CD40 ligand gene defects responsible for X-linked hyper-IgM syndrome. Science 259: 990

Armitage RJ, Fanslow WC, Strockbine L, Sato TA, Clifford KN, Macduff BM, Anderson DM, Gimpel SD, Davis-Smith T, Maliszewski CR, Clark EA, Smith CA, Grabstein KH, Cosman D, Spriggs MK (1992) Molecular and biological characterization of a murine ligand for CD40. Nature 357:80

Aruffo A, Farrington M, Hollenbaugh D, Li X, Milatovich A, Nonoyama S, Bajorath J, Grosmaire LS, Stenkamp R, Neubauer M, Roberts RL, Noelle RJ, Ledbetter JA, Francke U, Ochs HD (1993) The CD40 ligand, gp39, is defective in activated T cells from patients with X-linked hyper-IgM syndrome. Cell 72: 291

Banchereau J, Rousset F (1991) Growing human B lymphocytes in the CD40 system. Nature 353: 678

Banchereau J, de Paoli P, Vallé A, Garcia E, Rousset F (1991) Long term human B cell lines dependent on interleukin 4 and anti-CD40. Science 251: 70

Berek C (1992) The development of B cells and the B-cell repertoire in the microenvironment of the germinal center. Immunol Rev 126: 5–19

Berek C, Berger A, Apel M (1991) Maturation of the immune response in germinal centers. Cell 67: 1121

Bonnefoy JY, Henchoz S, Hardie D, Holder MJ, Gordon J (1993) A subset of anti-CD21 antibodies promote the rescue of germinal center B cells from apoptosis. Eur J Immunol 23: 969

Burton GF, Conrad DH, Szakal AK, Tew JG (1993) Follicular dendritic cells and B-cell costimulation. J Immunol 150: 31

Clark EA (1990) CD40: A cytokine receptor in search of a ligand. Tissue Antigens 35: 33

Clark EA, Ledbetter JA (1986) Activation of human B cells mediated through two distinct cell surface differentiation antigens, Bp35 and Bp50. Proc Natl Acad Sci USA 83: 4494

Cormann N, Lesage F, Heinen E, Schaaf-Lafontaine N, Kinet-Denoel C, Simar LJ (1986) Isolation of follicular dendritic cells from human tonsils and adenoids. V. Effect on lymphocyte proliferation and differentiation. Immunol Lett 14: 29

DiSanto JP, Bonnefoy JY, Gauchat JF, Fischer A, de Saint Basile G (1993) CD40 ligand mutations in X-linked immunodeficiency with hyper-IgM. Nature 361: 541

Foy TM, Laman JD, Ledbetter JA, Aruffo A, Claassen E, Noelle RJ (1994) gp39-CD40 interactions are essential for germinal center formation and the development of B cell memory. J Exp Med 180: 157

Freedman AS, Munro JM, Rice GE, Bevilacqua MP, Morimoto C, McIntyre BW, Rhynhart K, Pober JS, Nadler LM (1990) Adhesion of human B cells to germinal centers in vitro involves VLA-4 and INCAM-110. Science 249: 1030

Freedman AS, Munro JM, Rhynhart K, Schow P, Daley N, Lee N, Svahn J, Eliseo L and Nadler LM (1992) Follicular dendritic cells inhibit human B cell proliferation. Blood 80: 1284

Fuleihan R, Ramesh N, Loh R, Jabara H, Rosen FS, Chatila T, Fu S-M, Stamenkovic I, Geha RS (1993) Defective expression of the CD40 ligand in X chromosome-linked immunoglobulin deficiency with normal or elevated IgM. Proc Natl Acad Sci USA 90: 2170

Fuller KA, Kanagawa O, Nahm MH (1993) T cells within germinal centers are specific for the immunizing antigen. J Immunol 151: 4505

Gordon J, Millsum MJ, Guy GR, Ledbetter JA (1988) Resting B lymphocytes can be triggered directly through the CDw40 (Bp50) antigen. A comparison with IL-4 mediated signaling. J Immunol 140: 1425

Gray D (1988) Recruitment of virgin B cells into an immune response is restricted to activation outside lymphoid follicles. Immunology 65: 73

Gray D, Skarvall H (1988) B-cell memory is short-lived in the absence of antigen. Nature 336: 70

Hardie DL, Johnson GD, Khan M, MacLennan ICM (1993) Quantitative analysis of molecules which distinguish functional compartments within germinal centers. Eur J Immunol 23: 997

Jacob J, Kelsoe G (1992) In situ studies of the primary immune response to (4-hydroxy-3-nitrophenyl) acetyl. II. A common clonal origin for periarteriolar lymphoid sheath-associated foci and germinal centers. J Exp Med 176: 679

Jacob J, Kelsoe G, Rajewsky K, Weiss U (1991) Intraclonal generation of antibody mutants in germinal centers. Nature 354: 389

Kawabe T, Naka T, Yoshida K, Tanaka T, Fujiwara H, Suematsu S, Yoshida N, Kishimoto T, Kikutani H (1994) The immune response in CD40-deficient mice: impaired immunoglobulin class switching and germinal center formation. Immunity 1: 167

Koopman G, Parmentier HK, Schuurman H-J, Newman W, Meijer CJLM, Pals ST (1991) Adhesion of human B cells to follicular dendritic cells involves both the lymphocyte function-associated antigen 1/intercellular adhesion molecule 1 and very late antigen 4/vascular cell adhesion molecule 1 pathways. J Exp Med 173: 1297
Koopman G, Keehnen RM, Lindhout E, Newman W, Shimizu Y, Van Seventér GA, de Groot C, Pals ST (1994) Adhesion through the LFA-1 (CD11a/CD18)-ICAM-1 (CD54) and the VLA-4 (CD49d)-VCAM-1 (CD106) pathways prevents apoptosis of germinal center B cells. J Immunol 152: 3760
Korthäuer U, Graf D, Mages HW, Briére F, Padayachee M, Malcolm S, Ugazio AG, Notarangelo LD, Levinsky RJ, Kroczek RA (1993) Defective expression of T-cell CD40 ligand causes X-linked immunodeficiency with hyper-IgM. Nature 361: 539
Kosco MH (1991) Cellular interactions during the germinal centre response. Res Immunol 142: 245
Kosco MH, Gray D (1992) Signals involved in germinal center reactions. Immunol Rev 126: 63
Kosco MH, Burton GF, Kapasi ZF, Szakal AK, Tew JG (1989) Antibody-forming cell induction during an early phase of germinal centre development ant its delay with ageing. Immunology 68: 312
Kosco MH, Pflugfelder E, Gray D (1992) Follicular dendritic cell-dependent adhesion and proliferation of B cells in vitro. J Immunol 148: 2331
Kroese FGM, Wubbena AS, Seijen HG, Nieuwenhuis P (1987) Germinal centers develop oligoclonally. Eur J Immunol 17: 1069
Kroese, FGM, Timens, W and Nieuwenhuis, P (1990) Germinal center reaction and B lymphocytes: morphology and function. In: Grundmann E, Vollmer E (eds) Reaction patterns of the lymph node, part 1: cell types and functions. Springer, Berlin Heidelberg New York, p119 (Current topics in pathology, vol 84)
Küppers R, Zhao M, Hansmann M-L, Rajewsky K (1993) Tracing B cell development in human germinal centres by molecular analysis of single cells picked from histological sections. EMBO J 12: 4955
Ledbetter JA, Shu G, Gallagher M, Clark EA (1987) Augmentation of normal and malignant B cell proliferation by monoclonal antibody to the B cell-specific antigen BP50 (CDw40). J Immunol 138: 788
Lederman S, Yellin MJ, Inghirami G, Lee JJ, Knowles DM, Chess L (1992) Molecular interactions mediating T-B lymphocyte collaboration in human lymphoid follicles. Roles of T cell-B cell-activating molecule (5c8 antigen) and CD40 in contact-dependent help. J Immunol 149: 3817
Lindhout E, Mevissen ML, Kwekkeboom J, Tager JM, de-Groot C (1993) Direct evidence that human follicular dendritic cells (FDC) rescue germinal center cells from death by apoptosis. Clin Exp Immunol 91:330
Liu YJ, Joshua DE, Williams GT, Smith CA, Gordon J, MacLennan ICM (1989) Mechanism of antigen-driven selection in germinal centres. Nature 342: 929
Liu Y-J, Zhang J, Lane PJL, Chan EY-T, MacLennan ICM (1991) Sites of specific B cell activation in primary and secondary responses to T cell-dependent and T cell-independent antigens. Eur J Immunol 21: 2951
Liu Y-J, Johnson GD, Gordon J, MacLennan ICM (1992) Germinal centres in T-cell-dependent antibody responses. Immunol Today 13: 17
MacLennan ICM (1994) Germinal centers. Annu Rev Immunol 12: 117
McHeyzer-Williams MG, Nossal GJV, Lalor PA (1991) Molecular characterization of single memory B cells. Nature 350: 502
McHeyzer-Williams MG, McLean MJ, Lalor PA, Nossal GJV (1993) Antigen-driven B cell differentiation in vivo. J Exp Med 178: 295
Nossal GJV (1992) The molecular and cellular basis of affinity maturation in the antibody response. Cell 68: 1
Nossal GJV, Abbot A, Mitchell J, Lummus Z (1968) Antigens in immunity. Ultrastructural features of antigen capture in primary and secondary lymphoid follicles. J Exp Med 127: 277
Pascual V, Liu Y-J, Magalski A, de Bouteiller O, Banchereau J, Capra JD (1994) Analysis of somatic mutation in B-cell subsets of human tonsil correlates with phenotypic differentiation from the naive to the memory B-cell compartment. J Exp. Med 180: 329
Petrasch SG, Kosco MH, Perez-Alvarez CJ, Schmitz J, Brittinger G (1991) Proliferation of germinal center B lymphocytes in vitro by direct membrane contact with follicular dendritic cells. Immunobiology. 183: 451
Schittek B, Rajewsky K (1992) Natural occurence and origin of somatically mutated memory B cells in mice. J Exp Med 176: 427
Szakal AK, Hanna MG (1968) The ultrastructure of antigen localization and viruslike particles in mouse spleen germinal centers. Exp Mol Pathol 8: 75

Szakal AK, Gieringer RL, Kosco MH, Tew JG (1985) Isolated follicular dendritic cells: cytochemical antigen localization, Nomarski, SEM, and TEM morphology. J Immunol 134: 1349
Tew JG, Phipps RP, Mandel TE (1980) The maintenance and regulation of the humoral immune response: persisting antigen and the role of follicular antigen-binding dendritic cells as accessory cells. Immunol Rev 53: 175
Vallé A, Zuber CE, Defrance T, Djossou O, De Rie M, Banchereau J (1989) Activation of human B lymphocytes through CD40 and interleukin 4. Eur J Immunol 19: 1463
Van den Eertwegh AJM, Noelle RJ, Roy M, Shepherd DM, Aruffo A, Ledbetter JA, Bocrsma WJA, Claassen E (1993) In vivo CD40-gp39 interactions are essential for thymus dependent humoral immunity. I. In vivo expression of CD40 ligand, cytokines, and antibody production delineates sites of cognate T-B cell interactions. J Exp Med 178: 1555
Weiss U, Rajewsky K (1990) The repertoire of somatic antibody mutants accumulating in the memory compartment after primary immunization is restricted through affinity maturation and mirrors that expressed in the secondary response. J Exp Med 172: 1681
Weiss U, Zoebelein R, Rajewsky K (1992) Accumulation of somatic mutants in the B cell compartment after primary immunization with a T cell-dependent antigen. Eur J Immunol 22: 511

Follicular Dendritic Cells: Structure as Related to Function

K. Maeda, M. Matsuda, and Y. Imai

1 Introduction

Follicular dendritic cells (FDC) have been noted as one of unique immunological accessory cells. They are localized within lymphoid follicles as an intricate cellular meshwork that helps to maintain follicular structure. FDC also have the capability to trap and retain antigens in the form of immune complexes on their surface for long periods, providing unprocessed antigens to germinal center B cells in the form of iccosomes (immune complex-coated bodies). In this chapter, the morphological features and the cytochemical and immunological phenotypes of FDC are reviewed and discussed within the context of recent topics and controversial issues.

2 Microscopic Features: Light and Ultrastructural Levels

The anatomical localization, light microscopic morphology, ultrastructural features, and cytochemical properties of FDC are summarized in Table 1. FDC are characterized by their intricate, entangled, and arborizing cytoplasmic processes,

Department of Pathology, Yamagata University School of Medicine, 2-2-2 Iida-Nishi, Yamagata, Japan 990-23

Table 1. Morphological features of follicular dendritic cells (FDC)

	Feature	Reference
Anatomical localization	Within lymphoid follicles (primary and secondary follicles), especially in the light zone of germinal centers	LENNERT (1978) IMAI et al. (1983)
Light microscopic morphology	Nucleus: oval-shaped or elongated, often with irregular corners or angular; clearly defined nuclear membrane; a medium-sized prominent nucleolus; sometimes binucleated (26%) and rarely trinucleated Cytoplasm: weakly staining cytoplasm: No active phogocytic activity	LENNERT (1978) PETERS et al. (1984)
Ultrastructure	Nucleus: coarser chromatin, markedly condensed in a narrow zone along the inner nuclear membrane (euchromatic); central solitary nucleolus	NOSSAL et al.(1968)
	Cytoplasm: extremely narrow around the nucleus; well-developed smooth endoplasmic reticulum; no acid phosphatase-positive lysosomes; no phagosomes; very fine interwoven processes that form reticular network with other FDC; Desmosomal junctions connecting cell projections	LENNERT (1978) IMAI et al.(1983) RADEMAKERS (1992)
Cytochemistry	Nonspecific esterase: moderate (+), α–Naphthyl acetate esterase: moderate (+), 5'-Nucleotidase: strongly positive or negative (controversial reports)	MÜLLER-HERMELINK (1974)
	ATPase: weak or moderate (±)	LENNERT (1978)
	Acid phosphatase: negative (–) Alkaline phosphatase: negative (–)	HEINEN et al. (1985b)
	β– glucuronidase: negative (–) diphenylaminopeptidase IV: negative (–)	PARMENTIER et al. (1991)

ATPase, adenosine triphosphatase.

the frequent desmosomal junctions that connect the cell processes, and their euchromatic, angular nucleus. Sometimes they have a binucleated appearance. The frequent desmosomal junctions occurring between cell processes of FDC themselves or with adjacent germinal center cells demonstrate the intimate cellular contacts and functional adhesions FDC maintain with each other and with surrounding germinal center lymphocytes. The diversity of FDC morphology has recently been described by RADEMAKERS (1992), and seven different FDC subtypes have been proposed on the basis of ultrastructure (see SCHUURMAN et al., this volume).

In a suspension of isolated FDC, the above-mentioned features are also exhibited (LILET-LECLERCQ et al. 1984). Scanning electron microscope (EM) observations revealed distinctive morphology, as described by SZAKAL et al. (1985). Human FDC have been mainly isolated from tonsils. FDC appeared as cellular clusters enveloping several lymphocytes (six to 12 cells; TSUNODA and KOJIMA 1987; ENNAS et al. 1989). When FDC were isolated initially as single cells, they would readily form clusters with cocultured lymphocytes in vitro (TSUNODA et al. 1993).

Cytochemically, FDC do not reveal detectable acid and alkaline phosphatase or nonspecific esterase activities. FDC appear to exhibit relatively strong 5'-nucleotidase activity on their surface membrane, although this is somewhat controversial (MÜLLER-HERMELINK 1974; LENNERT 1978; HEINEN et al. 1985b;

Paramentier et al. 1991). These cytochemical features distinguish FDC from monocytes and macrophages.

3 Surface Molecule Distribution on Follicular Dendritic Cells

As summarized in Tables 2–7, the immunological phenotypes of human, rat, and murine FDC have been investigated in a number of studies. These accumulated results suggest and support certain biological properties and immunological behavior for FDC. However, there are still many discrepancies and controversial issues concerning the definitive functions of FDC.

3.1 Specific Markers for Follicular Dendritic Cells

Several monoclonal antibodies specific for human, rat, and murine FDC have been developed (Table 2). Unfortunately, most of these are not entirely specific and have cross-reactivity to some other types of cells. Among these, R4/23 and Ki-M4, ED5 and Ki-M4R, and FDC-M1 (Fig.1) are especially useful in defining FDC in human, rat, and mouse tissue, respectively.

3.2 Complement Receptors

It is generally accepted that FDC bear a large amount of complement receptors, especially CR1 (C3b receptor; CD35) and CR2 (C3d receptor; CD21) (Table 3). In addition, Schriever et al. (1991; Schriever and Nalder 1992) demonstrated using single cell polymerase chain reaction (PCR) techniques that FDC contained high levels of mRNA for CR2. These molecules are thought to involve the constitutive trapping and retention of immune complexes by FDC.

CR3 (C3bi receptor; CD11b) is known to be expressed on the surface of cells of the monocyte–macrophage lineage. Previous reports were conflicting as to the expression of CR3 by FDC. Recent observations using isolated cells from tonsils (Schriever et al. 1989; Sellheyer et al. 1989; Petrasch et al. 1990), however, confirmed that FDC actually bear CR3, at least in the human system (Table 3).

3.3 Fc Receptors

Previous reports, with one exception (Sellheyer et al. 1989), indicated that the low-affinity Fc receptor for immunoglobulin (Ig)E (FcεRII; CD23) was detectable on the surface of FDC (see Table 3). Interestingly, several reports mentioned that only a fraction of FDC (not all FDC) bore this molecule (Masuda et al. 1989;

Table 2. Monoclonal antibodies relatively specific for follicular dendritic cells (FDC)

Human FDC	Rat FDC	Mouse FDC
R4/23 (DRC-1) (NAIEM et al. 1983)	MRC OX-2 (MCMASTER and WILLIAMS 1979;BARKLAY 1981)	FDC-M1 (KOSCO et al. 1992; MAEDA et al. 1992b)
Cross-reaction with some B lymphocytes	Cross-reaction with endothelial cells, sinus lining cells, etc.	Cross-reaction with TBM and some endothelial cells
KiM4 (PARWARESCH et al. 1983) Cross-reaction with peroxidase–positive mononuclear cells in blood and U937 cells	ED5 (JEURISSEN and DIJKSTRA 1986) Cross-reaction with reticular fiber	–
BU -10 (JOHNSON et al. 1986)	KiM4R (WACKER et al. 1987) Cross–reaction with endothelial cells of HEV and sinus lining cells, etc.	–
Ki-FDC1p (TABRIZCHI et al. 1990)	ED11 (VAN DEN BERG et al. 1989) Recognizes C3	–
X-11 (WÜRZBER et al. 1991) Recognizes C9–related antigen	–	–
DF–DRC1 (manufacturer's instructions from Sera-Labo Inc. Sussex, England) Recognizes a 50-kDa molecule	–	–

TBM, tingible body macrophage; HEV, high endothelial venules.

PETRASCH et al. 1990), suggesting possible heterogeneity among FDC. In addition, our experimental data in mice using the monoclonal antibody B3B4, which recognizes the mouse equivalent of human CD23, indicated a clear expression by FDC at EM level (Fig. 2). This expression could be inducible by certain stimuli such as *Nippostrongylus brasiliensis* infection or exogenous antigenic challenges (MAEDA et al. 1992). Recently RIEBER et al. (1993) demonstrated that FDC could be induced by interleukin (IL)-4 to express membrane-bound FcεRII and to release a soluble form of this molecule. Also, FDC could be shown to contain mRNA for a selected isoform of FCεRII (CD23b). Although the actual functions of the molecule on FDC are still being investigated, our experiments suggested that this molecule be involved in the trapping and retention of IgE immune complexes by FDC and might possibly contribute certain IgE-mediated immune responses or allergic reactions (MAEDA et al. 1992a). This molecule is considered to be an important factor not only for binding to IgE or IgE immune complexes, but also for induction of B cell proliferation, differentiation, and rescue of germinal center B

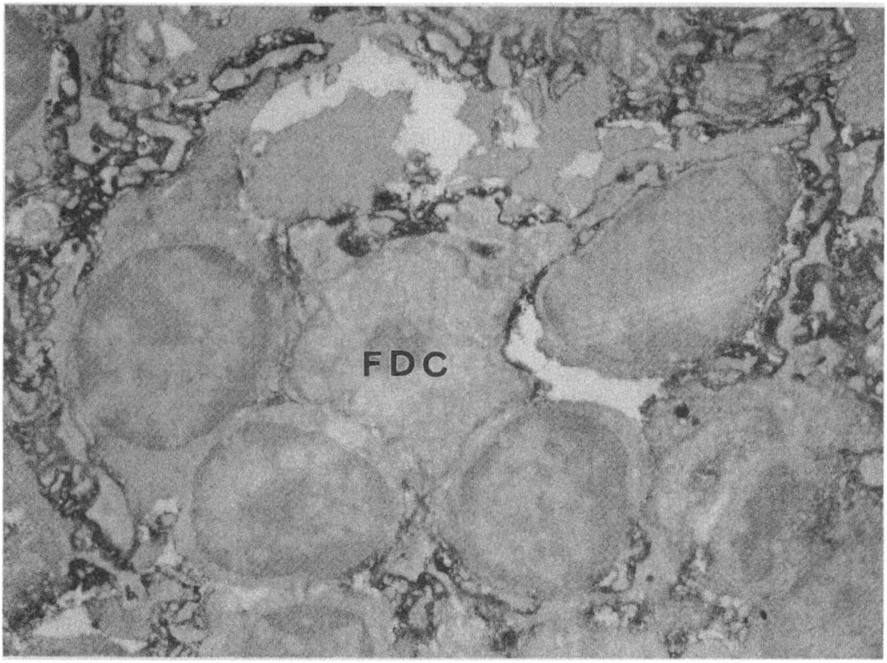

Fig. 1. An electron micrograph demonstrating the reactivity of FDC-M1 in the germinal center of actively immunized mouse (BALB/c) lymph node. A follicular dendritic cell (FDC) cell body is apparent with the typical euchromatic nucleus and dendritic processes emanating among surrounding lymphoid cells. The electron-dense (*black*) labeling on the plasma membrane of cell body and their cytoplasmic processes indicates the binding of FDC-M1. (Not counterstained, x 10 500)

cells from apoptosis. Future investigations on the regulation and kinetics of this molecule should elucidate the correlation of CD23 expression and its essential role in FDC in germinal centers.

In the murine system, it was demonstrated experimentally by Humphrey and Grennan (1982) and Radoux et al. (1985) that FDC bore Fc receptor on their surface. In addition, Kosco et al. (1986) and Maeda et al. (1992a, b) also reported that the monoclonal antibody 2.4G2, which recognizes murine low-affinity Fcγ receptors (FcγRII), labeled FDC very intensely in murine lymph nodes (LN). These observations were then confirmed at the EM level (Fig. 3). Recently, Yoshida et al. (1993) reported that 2.4G2 could partially inhibit the binding of immune complex to FDC in vitro, and they argue that FcγRII are partially involved in the trapping and retention of immune complexes by FDC.

In contrast, conflicting observations have been reported for human FDC regarding the expression of Fcγ receptors. Heinen et al. (1985) have reported that human tonsillar FDC expressed Fc receptors on their surface. In addition, Petrasch et al. (1990) have described a positive reaction between FDC isolated from human tonsils and lymphoma tissues and monoclonal antibodies directed against FcγRII (CDw32) and FcγRIII (CD16). In contrast, several reports involving

Table 3. Surface phenotype of follicular dendritic cells (FDC); complement receptors (CR), FC receptors (FcγR, FcεR), major histocompatibility complex (MHC) class I and II, and related molecules

Reference	Human FDC					
	GERDES et al. (1983)	VAN DER VALK et al. (1984)	HEINEN et al. (1984)	WOOD et al. (1985)	JOHNSON et al. (1986)	SCHRIEVER et al. (1989)
Source	Frozen tissues and suspended cells of tonsils and lymphoma tissues	Frozen tissues of LN, spleen, and tonsils	Isolated cells from tonsils	Frozen tissues of LN, skin, spleen, tonsils, liver, etc.	Frozen tissues of LN, skin, spleen, tonsils, lymphoma tissue, etc.	Isolated cells from tonsils
CR						
CR1 (CD35)	Pan C3R+++	Anti-CR1 serum ++	CR1 (To5)+++			CR1 (543)+
CR2 (CD21)	CR1 (To5) ++++				CR2 (BL13)++	CR2 (B2)+
CR3 (CD11a)	CR3(OKM-1)+/−	CR3(OKM-1)−				CR3 (Mo1)+
CD19					B4 ++	B4+
FcγR						
FcγRI(CD64)						CD16(Leu11b) −
FcγRII(CD32)						CD32(IV-3D3) −
FcγRIII(CD16)						
FcεR (CD23)					MHM6 + (subset)	Blast-2 +
MHC class I	W6/32HL ++			W6/32++		W6/32 +
MHC class II	TÜ22, TÜ35, TÜ39 (+)/+	++	TÜ35 +	L203 +		6/7 +

LN, Lymph nodes; PLP, periodate-lysine-paraformaldehyde solution for immunoelectron microscopy; EM, electron microscopy; LM, light microscopy.

			Mouse FDC	
Sellheyer et al. (1989)	Petrasch et al. (1990)	Parmentier et al. (1991)	Kosco et al. (1986)	Maeda et al. (1992a, b)
Isolated cells from tonsils	Isolated cells from tonsils, LN, and lymphoma tissues	Frozen tissues and isolated cells from tonsils	Isolated cells and frozen tissue of LN (LM and EM)	Frozen and PLP-fixed tissues of LN (LM and EM)
CR1 (To5)+++	CR1 (To5)+	CR1++		CR1 (8C12)+++ CR1/CR2 (7E9)+++ CR1/CR2 (7G6)+++
CR2 (2G7)+++	CR2 (B2)+			
CR3 (Mo1)++	CR3 (Mo1)+			
HD37–				
FcγRI (10.1)– FcγRII (IV.3, 2E1, 2, CIKM5)– FcγRIII (BW209/2, CLB-FcGranI, Leu11a, Leu11b)–	FcγRII (40kdFcR) + FcγRIII (Leu11b) +	FcγRI (32.2) ±/– FcγRII (IV.3) ±/– FcγRIII (gran 1) ±/–	FcγRII (2.4G2) +	FcγRII (2.4G2) +++
Blast-2, MHM6, TÜ1 –	Ki-B1+ (50% of FDC)			B3B4 ++
W6/32HL ++/+++			11–4.1 (H2-K)+	
TÜ35 ++/+++	HLA-DR +	J0103 +	MRCox4, 14–4.4, 10–2.16, MK-D6 +	I-A++

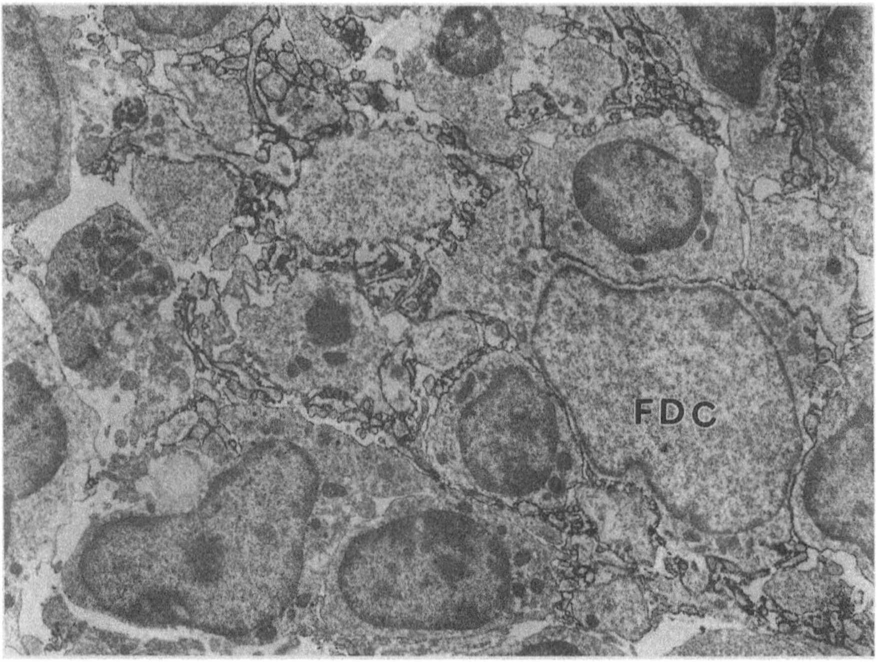

Fig. 2. An electron micrograph demonstrating the reactivity of monoclonal antibody B3B4, which recognize mouse FcεRII (CD23), in the germinal center of an actively immunized mouse's (C3H/He) lymph node. A cell body typical of follicular dendritic cells (*FDC*) and their intricately entangled cytoplasmic processes is clearly labeled with this antibody. (Not counterstained, x 8500)

the reactivity of FDC isolated from tonsils with a panel of antibodies could not confirm these results. They included monoclonal antibodies against FcγRI (CD64); FcγRII, and FcγRIII (SCHRIEVER et al. 1989; SELLHEYER et al. 1989; PARAMENTIER et al. 1991). These discrepancies might be explained by differences in isolation procedures of FDC, the antibodies employed, functional or biological phases of FDC, instability of the surface molecules, and/or heterogeneity of FDC. Further investigations, especially using molecular biological methods, may be able to resolve these inconsistencies, elucidating more precisely the mechanisms of antigenic capture by FDC.

3.4 Major Histocompatibility Complex Class I and II Molecules

Whereas all reports concurred that FDC express MHC class I molecules on the surface, their expression of class II MHC molecules was controversial. HUMPHREY and GRENNAN (1982) reported that FDC did not have MHC class II molecules. In contrast, many authors have described the presence of these molecules on FDC in both human and mouse tissues (GERDES et al. 1983; SELLHEYER et al. 1989;

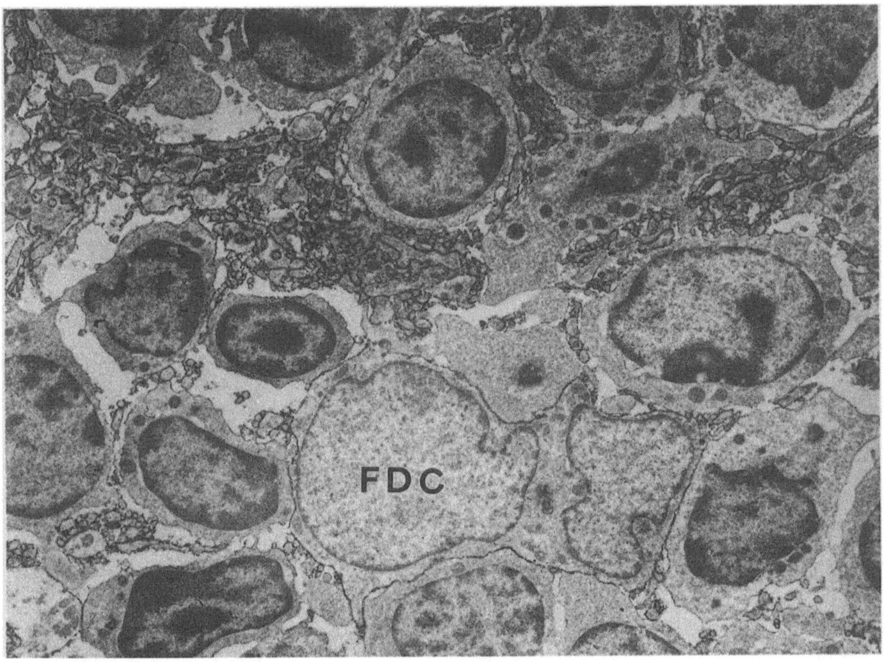

Fig. 3. An electron micrograph demonstrating the reactivity of 2.4G2, which recognize mouse FcγRII, in the germinal center of an actively immunized mouse's (C3H/He) lymph node. A follicular dendritic cell (FDC) cell body is apparent with the characteristic binucleated euchromatic nucleus and dendritic processes entangled intricately. The electron-dense (*black*) labeling on the plasma membrane of the cell body and the cytoplasmic processes indicates the binding of 2.4G2. (Not counterstained, x 8500)

SCHRIEVER et al. 1989; PETRASCH et al. 1990; KOSCO et al. 1986; MAEDA et al. 1992; see Table 3). Recently, GRAY et al. (1991) offered new explanations for this discrepancy on the basis of experiments using chimeric mice. They concluded that FDC did not synthesize MHC class II, but acquired MHC molecules shed from B lymphocytes. Thus far it has not been proved that FDC present antigens to T cells directly in vivo. However, MHC class II molecules might play a key role in the cellular relationship between FDC and T cells.

3.5 Immunoglobulins and Complement Components

Most of the reports concerning the immunoglobulins present on FDC were consistent in their observation that FDC bore all isotypes of immunoglobulins except for IgD (see Table 4). These immunoglobulins are regarded as a constituent of the immune complexes trapped and retained by FDC. IgM was especially detected in most cases, and this immunoglobulin isotype might play a significant role in eliciting germinal center reactions. Our preliminary examination using monoclonal antibodies against subclasses of human IgG and IgA

Table 4. Surface phenotype of follicular dendritic cells (FDC) on the immunoglobulins (Ig) and complement components

Reference	Human FDC				
	GERDES et al. (1983)	VAN DER VALK et al. (1984)	HEINEN et al. (1984)	IMAI et al. (1986)	HALSTENSEN et al. (1988)
Source	Frozen tissues and cell suspension of tonsils and lymphoma tissues	Frozen tissues of LN, spleen, and tonsils	Isolated cells from tonsils	Frozen tissues of LN, thyroid, synovium, etc.	Frozen tissue of LN, spleen, and tonsils
IgG	++	+	++ or +/−	++	
IgM	+++	++	++	+++	
IgA	++		++ or −	+	
IgD	−		−	−	
IgE				− or +++	
Light chains	κ +++ λ +++	κ + λ +			
Complement Components	C3b ++			C1q ++, C3c +, C3d +++, C4 +, C5+, C6−, C8−, C9−, properdin +, C3act −, C3bINA +, β1H−	C3d + C5 + TCC + S−protein +

LN, Lymph nodes; PLP, periodate-lysine-paraformaldehyde solution for immunoelectron microscopy; TCC, terminal complement complex; EM, electron microscopy; LM, light microscope.

				Rat FDC	Mouse FDC
SCHRIEVER et al. (1989)	SELLHEYER et al. (1989)	ZWIRNER et al. (1989)	PARMENTIER et al. (1991)	VAN DEN BERG et al. (1989)	MAEDA et al. (1992a, b)
Isolated cells from tonsils	Isolated cells from tonsils	Frozen tissues of LN, thymus, and tonsils	Frozen tissues and isolated cells from tonsils	Frozen tissues of LN, spleen, thymus, Peyer's patch, etc.	Frozen and PLP-fixed tissues of LN (LM and EM)
−	+++	+/−	++		IgG1++, IgG2a++, IgG2b++, IgG3++,
+	++/+++	+++	++		++
−	++/+++				++/+
−	−		±/−		−
−					++/−
κ + λ −					
		C1q +/±, C1r±, C1s±, C2−, C3d +++, C3c +/−, C4a−, C4c ±/−, C4d +++, C5 ++/−, C9 +/−, C4bp ++/−, factorB −, factorH −, properdin −	C1q ++, C3b +	C3 (ED11) ++	

Table 5. Surface phenotypes of follicular dendritic cells (FDC) on the molecules restricted to B and T lymphocytes

Reference	Human FDC				
	GERDES et al. (1983)	VAN DER VALK et al. (1984)	HEINEN et al. (1984)	WOOD et al. (1986)	CARBONE et al. (1988)
Source	Frozen tissue and cell suspensions of tonsils and lymohoma tissues	Frozen tissues of LN, spleen, and tonsils	Isoloated cells from tonsils	Frozen tissues of LN, spleen, tonsils skin, etc.	Frozen tissues of LN, tonsils, and spleen
CD1	OKT6 – NA1/34 –	OKT6 – NA1/34 –		Leu6 –	
CD2	OKT11 –				
CD3					
CD4	OKT4 –	Leu3a –		Leu3a +	
CD5	Leu1 –	Leu1 –			
CD8	OKT8 –	Leu2a –			
Other T cell markers		TA1 –	Tü1 + or –		
CD9		BA–2 +			BA–2+/–
CD10	VIL-A1 +/–				
CD20					
CD22			To15/54 –		
CD24		BA–1 –			
CD25					
CD30					
CD37					
CD38					
CD40					
CD45 (LCA)					
CD45R					
Other B cell markers				L3B12 ++	
CD56					
CD71					

LN, Lymph nodes; PLP, periodate-lysine-paraformaldehyde solution for immunoelectron microscopy; LCA, leukocyte common antigen; EM, electron microscopy; LM, light microscopy.

				Mouse FDC	
SCHRIEVER et al. (1989)	SELLHEYER et al. (1989)	PETRASH et al. (1989)	PARMENTIER et al. (1991)	KOSCO et al. (1989)	MAEDA et al. (1992a, b)
Isolated cells from tonsils	Isolated cells from tonsils,	Isolated cells from LN, tonsils, and lymphoma tissues	Frozen tissues and isolated cells from tonsils	Isolated cells and frozen tissues of LN (LM and EM)	Frozen and PLP-fixed tissues of LN (LM and EM)
CD1a(T6) – CD1b(T009) – CD1c(T024) –			OKT6 –		
T11 –			T11–		
T3 –			Leu4 –		
T4 –	OKT4 – T310 –		Leu3a ± ADP336 ± ADP302 ± ADP359 ±		L3T4 –
T1 –			Leu1 –	Ly1.1– Ly1.2 –	Ly1 –
T8 –	Tü102 –		Leu2a –	Ly2.1 –	Lyt2 –
CD6 (2H1) – CD7 (T159) –				Thy1.2 –	
J2 –					
J5 –					
B1 –		B1+			
G28–7 – HD39 –			To15 ++		
HB8 –		OKB2 + (50% of FDC)			
IL-2R –			J0101–		
RSC1 –			Ki–1 –		
G28–1 +			IOB–1 ±		
		OKT10 +			
G28–5 +					
T200 –		2B11 & PD7/26 +		M1/9.3 +	T200 –
2H4 –					B220 –
B5 –, B7 –, PCA-1 –					
NKH-1 –					
		L01.1 +			

Table 6. Phenotype of follicular dendritic cells of (FDC) on the molecules restricted to myelo-macrophages and related cells

Reference	Human FDC				
	GERDES et al. (1983)	VAN DER VALK et al. (1984)	HEINEN et al. (1984)	WOOD et al. (1985)	CARBONE et al. (1986)
Source	Frozen tissue and cell suspensions of tonsils and lymphoma tissues	Frozen tissues of LN, spleen, and tonsils	Isoloated cells from tonsils	Frozen tissues of LN, spleen, tonsils, skin, liver, etc.	Formalin-fixed, paraffin embedded tissues of lymphoma
CD11b					
CD11c					
CD14			– or +	++	
CD13					
CD15	TÜ9 +/–				
CD33					
CD34					
CD38					
CD39					
Lysozyme	–				
α_1–AT	++				
α_1–ACT	+				
S–100 protein				–	+ S–100α ++ S–100β +/– (TANAKA 1986)
Others	Antimonocyte-1 ++ Antimonocyte-2 + TÜ5 –, TÜ6 –	Antihuman monocyte-1 +		63D3 ++ 61D3 ++	

α_1–AT, α_1-antitripsin; α_1–ACT, α_1–antichymotripsin; LN, lymph nodes; PLP, periodate-lysine-paraformaldehyde solution for immunoelectron microscopy; EM, electron microscopy; LM, light microscopy.

				Rat FDC	Mouse FDC
SCHRIEVER et al. (1989)	SELLHEYER et al. (1989)	PETRASCH et al. (1990)	PARMENTIER et al. (1991)	VAN DEN BERG et al. (1989)	HUMPHERY and GRENNAN (1982); KOSCO et al. (1986); MAEDA et al. (1992a, b)
Isolated cells from tonsils	Isolated cells from tonsils	Isolated cells from tonsils LN and lymphoma tissues	Frozen tissues and isolated cells from tonsils	Frozen tissues of LN, thymus, spleen, etc.	Isolated cells and frozen and PLP-fixed tissues of LN (LM and EM)
		Mo1 +			Mac-1 (M1/70)
–	SHCI3 –				
Mo2 +	MEM15 ++/+++	LeuM3 +			
My7 –					
	LeuM1 –				
My9 –					
12–8 –					
5D2 –					
G28-8 –					
	–				
	–		–	++ (COCCHIA et al. 1983)	
63D3+	Ber-MAC3 – MAC387 –		EBM11 ±	ED10 – ED11 ++ ED12 – ED13 – ED14 – ED15 –	Mac–2 (M3/38) – Mac–3 (M3/84) – F4/80 – MAS 034 –

Table 7. Surface phenotypes of follicular dendritic cells (FDC) on the molecules involving cellular adhesion

	Human FDC						Mouse FDC
	CARBONE et al. (1987, 1988)	HALSTENSEN et al. (1988)	SCHRIEVER et al. (1989)	FREEDMAN et al. (1990)	KOOPMAN et al. (1991)	ZUTTER (1991)	MAEDA et al. (1992b)
Source	Frozen tissues of LN	Frozen tissues of LNs, tonsils,	Isolated cells of LN, spleen,	Frozen tissues from tonsils spleen, etc	Frozen tissues of tonils and tonsils	Frozen tissues and isolated cells	Frozen and PLP-fixed tissues of tonsils, LN, spleen, etc. from tonsils
Observation	LM	LM	LM	Binding assay	LM	LM	LM and EM
CD11α (LFA1α)			2F12 –		CLB–LFA–1/2 –		
CD18 (β–chain of CD11)			10F12 –		CLB–LFA–1/1 –		
CD29 (β_1–integrin)			4B4 +			4B4 ++++	
CD 41 (gpIIb/IIa)			–				
CD44					NKI–P2 –		
CD49a (VLA–1)			TS –			Ts2/7 +++	
CD49b (VLA–2))						12F1, P1E6 +/++	
CD49c (VLA–3)			J143 +			P1B5++/+++	
CD49d (VLA–4)			8F2 +		HP1/3, HP2/1 –	B-5G10 ++++	
CD49e (VLA–5)			B1E5 +			P1D6 +++	
CD49f (VLA–6)			GOH3 +			G0H3 ++++	
CD54 (ICAM–1)			RR1 +	Minimal inhibition	F10.2, RR1/1 ++		MK-1 ++
CD56 (NCAM–1)			NKH–1 –				
CD61(β_3–intergin)						7E3 –	
CD62E (ELAM–1)					7A9 –		
CD106 (VCAM–1 or INCAM–110)				++	2G7, 4B9 ++		
Desmoplakin-1 and –2	+						
S–protein (vitronectin)		+					
Factor VIII			–				
EMA			–				

LN, Lymph nodes; LM, light microscopy; PLP, periodate-lysine-paraformaldehyde solution for immunoelectron microscopy; EM electron microscopy; LFA, lymphocyte function-associated antigen; gp glycoprotein; VLA, very late antigen; ICAM, intercellular adhesion molecule; ELAM, endothelial leukocyte adhesion molecule; INCA, inducible cell adhesion molecule; NCAM, neural cell adhesion molecule; EMA, epithelial membrane antigen.

demonstrated the presence of IgG_2, IgG_3, IgG_4, and IgA_1 on FDC, but no significant presence of IgG_1 and IgA_2. In addition, distinctive retention of IgA immune complexes on FDC in tonsils of IgA nephropathy patients and IgE immune complexes in the regional lymph nodes of Kimura's disease (eosinophilic lymphofolliculoid granuloma with hyper-IgE) have been reported (KUSAKARI et al. 1994; DEGAWA et al. 1994; MASUDA et al. 1989) and the presence of these particular immune complex isotypes associated with FDC suggests the pathological or pathogenetical roles of FDC in the diseases.

Several reports have indicated that components of the complement system are present in the germinal centers of human lymphoid tissues. With little variations, almost all complement components have been detected in association with FDC. These components are considered to also be components of the immune complexes trapped and retained by FDC. An interesting question is, then, why in the presence of these activated complement components FDC and germinal center structures are not destroyed. The answer is still unresolved, but it has been suggested that S-protein (vitronectin) or decay accelerating factor (DAF; CD55), which have been reported to be present on FDC, might protect the cells from complement-mediated cellular destruction (HALSTENSEN et al. 1988; LAMPERT et al. 1993).

3.6 Molecules Restricted to T Lymphocytes

As shown in Table 5, FDC did not express most T cell-restricted markers. Some authors have described the expression of CD4 by human FDC (WOOD et al. 1985; CARBONE et al. 1988). CD4 expression by FDC has been given special attention in relation to the peculiar follicular lesions in human immunodeficiency virus (HIV) infections and related lymphadenopathies. However, other groups were not able to confirm the expression of this molecule in human or mouse system.

3.7 Molecules Restricted to B Lymphocytes

FDC share some similar molecules thought to be exclusively distributed on B cells, especially CD20, CD22, CD24, CD45, and L3B12 (see Table 5). These observations, however, are not confirmed by all, and most of them are controversial. Immunohistochemically or immunocytologically evaluating cell markers on FDC is quite difficult because FDC are so closely associated with B cells, even in single cell suspensions. Thus, caution is required in making a final decision.

3.8 Molecules Restricted to Myelo-Macrophages

As shown in Table 6, FDC have been reported to share some antigens distributed on the cells of the myelo-macrophage lineages. Several reports confirmed that FDC bore CD14 and 63D3 on their surface. Some authors argued that FDC may be possibly derived from myelo-macrophages or related cells based on these observations (GERDES et al. 1983; PARWARESCH et al. 1983). Most of these molecules, however, are not perfectly restricted to myelo-macrophages and are known to be distributed on a variety of cell lineages. Consequently, further molecular or experimental evidence will be necessary to define the cellular origin of FDC.

In rat, FDC react intensely with anti-S-100 protein (COCCHIA et al. 1983; IMAI et al. 1991). In addition, some reports indicated human FDC are also positive for the antibodies, especially for monoclonal antibody against S-100α subunit (CARBONE et al. 1986; TANAKA 1986). Our observations confirmed these results. In contrast, murine FDC did not react with similar antibodies, even though peripheral nerve bundles were positive in the tissues (MAEDA et al. 1992). Investigating the different reactivity of FDC to S-100 protein among species is quite interesting phylogenically, and such approaches may contribute to resolving the problems concerning the origin and life cycle of FDC.

3.9 Molecules Involving Cellular Adhesion

Recently, extensive investigations have been performed to elucidate the distribution of adhesion molecules on FDC. They have contact with numerous cells, and the cellular adhesions between FDC and adjacent cells should be an essential component of the cellular interaction within germinal centers. Thus far, FDC have been shown to express many adhesion molecules on their surfaces (see Table 7). In particular, intercellular adhesion molecule (ICAM)-1 (CD54; ligand for lymphocyte function-associated antigen-1, LFA-1) and vascular cell adhesion molecule (VCAM)-1 (CD106: inducible cell adhesion molecule, INCAM-110, ligand for very late antigen-4, VLA-4) were found to be important for cell–cell interactions.

Finally, precise and careful analysis of the phenotype of FDC has provided significant insights into the molecular mechanisms of FDC functions, the cellular relationships of FDC and other cell types, their cellular origin, and pathological behavior. Further analysis will be needed to completely understand FDC biology; we hope this review will be helpful for future approaches involving the analysis of FDC.

Acknowledgment. The authors wish to thank Dr. Marie H. Kosco-Vilbois at the Basel Institute for Immunology for her review of our manuscript and her helpful suggestions.

References

Barklay AN (1981) Different reticular elements in rat lymphoid tissue identified by localization of Ia, Thy-1 and MRC OX-2 antigens. Immunology 44: 727–736

Carbone A, Manconi R, Poletti A, Colombatti A, Tirelli U, Volpe R (1986) S-100 protein, fibronectin and laminin immunostaining in lymphomas of follicular center cell origin. Cancer 58: 2169–2176

Carbone A, Poletti A, Manconi R, Gloghini A, Volpe R (1987) Heterogeneous in situ immunophenotyping of follicular dendritic reticulum cells in malignant lymphomas of B-cell origin. Cancer 60: 2919–2926
Carbone A, Poletti A, Manconi R, Gloghini A, Volpe R (1988) Heterogeneous immunostaining patterns of follicular dendritic reticulum cells in human lymphoid tissue with selected antibodies reactive with differenct cell lineages. Hum Pathol 19: 51–56
Chen LL, Adams JC, Steinman RM (1978) Anatomy of germinal centers in mouse spleen, with special reference to "follicular dendritic cells." J Cell Biol 77: 148–164
Cocchia D, Tiberio G, Santarelli R, Michetti F (1983) S-100 protein in "follicular dendritic" cells of rat lymphoid organs. An immunohistochemical and immunocytochemical study. Cell Tissue Res. 230: 95–103
Degawa N, Maeda K, Nagashima R, Matsuda M, Arai S, Imai Y (1994) Immunohistochemical analysis on tonsillar tissues of the patients with IgA nephropathy. I Localization of IgA and the related molecules. Dendritic Cells 4: 101–111
Dijkstra CD, Van den Berg TK (1991) The follicular dendritic cell: possible regulatory roles of associated molecules. Res Immunol 142: 227–231
Ennas MG, Chilosi M, Scarpa A, Lantini MS, Cadeddu G, Fiore-Donati L (1989) Isolation of multicellular complexes of follicular dendritic cells and lymphocytes: immunophenotypical characterization, electron microscopy and culture studies. Cell Tissue Res 257: 9–15
Freedman AS, Munro JM, Rice GE, Bevilacqua MP, Morimoto C, McIntyre BW, Rhynhart K, Pober JS, Nadler LM (1990) Adhesion of human B cells to germinal centers in vitro involves VLA-4 and INCAM-110. Science 249: 1030–1033
Gerdes J, Stein H, Mason DY, Ziegler A (1983) Human dendritic reticulum cells of lymphoid follicles: their antigenic profile and their identification as multinucleated giant cells. Virchow Arch [B] 42: 161–172
Gray D, Kosco M, Stockinger B (1991) Novel pathway of antigen presentation for the maintenance of memory. Int Immunol 3: 141–148
Halstensen TS, Mollnes TE, Brandtzaeg P (1988) Terminal complement complex (TCC) and S-protein (vitronectin) on follicular dendritic cells in human lymphoid tissues. Immunology 65: 193–197
Heinen E, Lilet-Leclercq C, Mason Dy, Stein H, Boniver J, Radoux D, Kinet-Denoël C, Simar LJ (1984) Isolation of follicular dendritic cells from human tonsils and adenoids. II. Immunocytochemical characterization. Eur J Immunol 14: 267–273
Heinen E, Radoux D, Kinet-Denoël C, Moeremans M, De Mey J, Simar LJ (1985a) Isolation of follicular dendritic cells from human tonsils and adenoids. III. Analysis of their Fc receptors. Immunology 54: 777–784
Heinen E, Kinet-Denoël C, Simar LJ (1985b) 5′-Nucleotidase activity in isolated follicular dendritic cells. Immunol Lett 9: 75–80
Humphrey JH, Grennan D (1982) Isolation and properties of spleen follicular dendritic cells. In: Nieuwenhuis, van den Broek, Hanna (eds) In vivo immunology. Plenum, New York, pp 823–827
Imai Y, Terashima K, Matsuda M, Dobashi M, Maeda K, Kasajima K (1983) Reticulum cell and dendritic reticulum cell -origin and function-. Rec Adv RES Res 21: 51–81
Imai Y, Yamakawa M, Masuda A, Sato T, Kasajima T (1986) Function of the follicular dendritic cell in the germinal center of lymphoid follicles. Histol Histophathol 1: 341–353
Imai Y, Matsuda M, Maeda K, Yamakawa M, Dobashi M, Satoh H, Terashima K (1991) Dendritic cells in lymphoid tissues—their morphology, antigenic profiles and functions. In: Imai Y, Tew JG, Hoefsmit ECM (eds) Dendritic cells in lymphoid tissues. Excepta Medica, Amsterdam, pp 3–13
Jeurissen SHM, Dijkstra CD (1986) Characteristics and functional aspects of nonlymphoid cells in rat germinal centers, recognized by two monoclonal antibodies ED5 and ED6. Eur J Immunol 16: 562–568
Johnson GD, Hardie DL, Ling NR, MacLennan ICM (1986) Human follicular dendritic cells (FDC): a study with monoclonal antibodies (MoAb). Clin Exp Immunol 64: 205–213
Koopman G, Parmentier HK, Schuurman H-J, Newman W, Meijer CJLM, Pals ST (1991) Adhesion of human B cells to follicular dendritic cells involves both the lymphocyte function-associated antigen 1/intercellular adhesion molecule 1 and very late antigen 4/vascular cell adhesion molecule 1 pathways. J Exp Med 173: 1297–1304
Kosco MH, Tew JG, Szakal AK (1986) Antigenic phenotyping of isolated and in situ rodent follicular dendritic cells (FDC) with emphasis on the ultrastructural demonstration of Ia antigens. Anat Rec 215: 201–213
Kosco MH, Pflugfelder E, Gray D (1992) Follicular dendritic cell-dependent adhesion and proliferation of B cells in vitro. J Immunol 148: 2331–2339

Kusakari C, Nose M, Takasaka T, Yuasa R, Kato M, Miyazono K, Fujita T, Kyogoku M (1994) Immunopathological features of palantine tonsil characteristic of IgA nephropathy; IgA1 localization in follicular dendritic cells. Clin Exp Immunol 95: 42–48

Lampert IA, Schofield JB, Amlot P, Van Noorden S (1993) Protection of germinal centers from complement attack: decay-accelerating factor (DAF) is a constitutitve protein on follicular dendritic cells. A study in reactive and neoplastic follicles. J Pathol 170: 115–120

Lennert K (1978) Dendritic reticulum cell (non-branching non-phagocytosing reticulum cell) In: malignant lymphomas other than Hodgkin's disease. Histology, cytology, ultrastructure immunology. Springer, Berlin Heidelberg New York, pp 59–65

Lilet-Leclercq C, Radoux D, Heinen E, Kinet-Denoël C, Defraige J-O, Houben-Defresne M-P, Simar LJ (1984) Isolation of follicular dendritic cells from human tonsils and adenoids. I. Procedure and morphological characterization. J Immunol Methods 66: 235–244

Maeda K, Burton GF, Padgett DA, Conrad DH, Huff TF, Masuda A, Szakal AK (1992a) Murine follicular dendritic cells and low affinity Fc receptors for IgE (FcεRII). J Immunol 148: 2340–2347

Maeda K, Matsuda M, Imai Y, Szakal AK, Tew JG (1992b) An immunohistochemical study of the phenotype of murine follicular dendritic cells (FDC). Dendritic Cells 1: 23–30

McMaster WR, Williams AF (1979) Identification of Ia glycoproteins in rat thymus and purification from rat spleen. Eur J Immunol 9: 426–433

Masuda A, Kasajima T, Mori N, Oka K (1989) Immunohistochemical study of low affinity Fc receptor for IgE in reactive and neoplastic follicles. Clin Immunol Immunopathol 53: 309–320

Müller-Hermelink HK (1974) Characterization of the B-cell and T-cell region of human lymphatic tissue through enzyme histochemical demonstration of ATPase and 5′-nucleotidase activities. Virchow Arch [B] 16: 371–378

Naiem M, Gerdes H, Abdulaziz Z, Stein H, Mason DY (1983) Production of a monoclonal antibody reactive with human dendritic reticulum cells and its use in the immunohistological analysis of lymphoid tissue. J Clin Pathol 36: 167–175

Nossal GJV, Abbot A, Mitchell J, Lummus Z (1968) Antigen in immunity. XV. Ultrastructural features of antigen capture in primary and secondary lymphoid follicles. J Exp Med 127: 277–290

Parmentier HK, Van der Linden JA, Krijnen J, Van Wichen DF, Rademakers LHPM, Bloem AC, Schuurman H-J (1991) Human follicular dendrtic cells: isolation and characteristics in situ and in suspension. Scand J Immunol 33: 441–452

Parwaresch MR, Radzun HJ, Hansmann M-L, Peters K-P (1983) Monoclonal antibody Ki-M4 specifically recognizes human dendritic reticulum cells (follicular dendritic cells) and their possible precursor in blood. Blood 62: 585–590

Peters JPJ, Rademakers LHPM, Roelofs JMM, de Jong D, van Unnik JAM (1984) Distribution of dendritic reticulum cells in follicular lymphoma and reactive hyperplasia. Light microscopic identification and general morphology. Virchow Arch [B] 46: 215–228

Petrasch S, Perez-Alvarez C, Schmitz J, Kosco M, Brittinger G (1990) Antigenic phenotyping of human follicular dendritic cells isolated from nonmalignant and malignant lymphatic tissue. Eur J Immunol 20: 1013–1018

Rademakers LHPM (1992) Dark and light zones of germinal centers of the human tonsil: an ultrastructural study with emphasis on heterogeneity of follicular dendritic cells. Cell Tissue Res 269: 359–368

Radoux D, Kinet -Denoël C, Heinen E, Moeremans M, De Mey J, Simar LJ (1985) Retention of immune complexes by Fc receptors on mouse follicular dendritic cells. Scand J Immunol 21: 345–353

Rieber EP, Rank G, Köhler I, Krauss S (1993) Membrane expression of FceRII/CD23 and release of soluble CD23 by follicular dendritic cells. Adv Exp Med Biol 329: 393–398

Schriever F, Nadler LM (1992) The central role of follicular dendritic cells in lymphoid tissues. Adv Immunol 51: 243–284

Schriever F, Freedman AS, Freeman G, Massner E, Lee G, Daley J, Nadler LM (1989) Isolated human follicular dendritic cells display a unique antigenic phenotype. J Exp Med 169: 2043–2058

Schriever F, Freeman G, Nadler LM (1991) Follicular dendritic cells contain a unique gene repertoire demonstrated by single-cell polymerase chain reaction. Blood 77: 787–791

Sellheyer K, Schwarting R, Stein H (1989) Isolation and antigenic profile of follicular dendritic cells. Clin Exp Immunol 78: 431–436

Szakal AK, Gieringer RL, Kosco MH, Tew JG (1985) Isolated follicular dendritic cells: cytochemical antigen localization, Nomarski, SEM, and TEM morphology. J Immunol 134: 1349–1359

Tabrizchi H, Hansmann M-L, Parwaresch MR, Lennert K (1990) Distribution pattern of follicular dendritic cells in low grade B-cell lymphomas of the gastrointestinal tract immunostained by Ki-FDC1p: a new paraffin-resistant monoclonal antibody. Mod Pathol 3: 470–478

Tanaka Y (1986) Immunocytochemical study of human lymphoid tissues with monoclonal antibodies against S-100 protein subunits. Virchows Arch [A] 410: 125–132

Tew JG, Thorbecke J, Steinman RM (1982) Dendritic cells in the immune response: characteristics and recommended nomenclature (a report from the reticuloendothelial society committee on nomenclature). J RES 31: 371–380

Tsunoda R, Kojima M (1987) A light microscopical study of isolated follicular dendritic cell-clusters in human tonsil. Acta Pathol Jpn 37: 575–585

Tsunoda R, Nakayama M, Heinen E, Miyake K, Suzuki K, Okamura H, Sugai N, Kojima M (1993) Emperipholesis of lymphoid cells by human follicular dendritic cells in vitro. Adv Exp Med Biol 329: 365–370

Van den Berg TK, Dopp EA, Breve JJP, Kraal G, Dijkstra CD (1989) The heterogeneity of the reticulum of rat peripheral lymphoid organs identified by monoclonal antibodies. Eur J Immunol 19: 1747–1756

van der Valk P, van der Loo EM, Jansen J, Daha MR, Meijer CJLM (1984) Analysis of lymphoid and dendritic cells in human lymph node, tonsil and spleen. A study using monoclonal and heterogeneous antibodies. Virchow Arch [B] 45: 169–185

Wacker H-H, Radzun HJ, Mielke V, Parwaresch MR (1987) Selective recognition of rat follicular dendritic cells (dendritic reticulum cells) by a new monoclonal antibody Ki-M4R in vitro and in vivo. J Leukoc Biol 41: 70–77

Wood GS, Turner RR, Shiurba RA, Eng L, Warnke RA (1985) Human dendritic cells and macrophages. In situ immunophenotypic definition of subsets that exhibit specific morphologic and microenvironmental characterisitics. Am J pathol 119: 73–82

Würzner R, Xu H, Franzke M, Schulze M, Peters JH, Götze O (1991) Blood dendritic cells carry terminal complement complexes on their cell surface as detected by newly developed neoepitope-specific monoclonal antibodies. Immunology 74: 132–138

Yamakawa M, Takagi M, Tajima K, Ohe S, Osanai T, Kudo S, Ito M, Imai Y (1991) Localization of blood coagulation factors and fibrinolysis factors within lymphoid germinal centers in human lymph nodes. Histochemistry 96: 123–127

Yoshida K, Van den Berg TK, Dijkstra CD (1993) Two functionally different follicular dendritic cells in secondary lymphoid follicles of mouse spleen, as revealed by CR1/2 and FcγRII-mediated immune-complex trapping. Immunology 80: 34–39

Zutter MM (1991) Immunolocalization of integrin receptors in normal lymphoid tissues. Blood 77: 2231–2236

Zwirner J, Felber E, Schmidt P, Riethmüller G, Feucht HE (1989) Complement activation in human lymphoid germinal centers. Immunology 66: 270–277

Follicular Dendritic Cells Initiate and Maintain Infection of the Germinal Centers by Human Immunodeficiency Virus

K. TENNER-RACZ and P. RACZ

1 Introduction

Since the discovery of the human immunodeficiency virus type 1 (HIV-1) as the cause of the acquired immunodeficiency syndrome (AIDS), careful attention has been paid to the preference of the virus for specific cells. However, the preference of HIV-1 for specific host tissues or tissue compartments such as germinal centers (GC) of the lymphoid tissue was often overlooked. Application of electronmicroscopic, immunohistochemical, and nucleic acid hybridization techniques to the study of lymphoid tissues led to the demonstration that GC are infected by HIV-1 and improved our understanding of the pathogenesis of AIDS.

The concept that in the disease caused by HIV-1 GC represent permanently infected foci in which HIV-1 persists even during clinical latency (TENNER-RACZ et al. 1986, 1988a) has slowly gained widespread acceptance. Recent studies with highly sensitive techniques such as DNA and RNA polymerase chain reaction (PCR) clearly show that peripheral blood insufficiently reflects the virological events during HIV infection. Simultaneous analyses of $CD4^+$ lymphocytes from blood and lymph nodes by DNA PCR demonstrated that the infection rate is of higher magnitude in the node than in the circulation (SCHNITTMAN et al. 1991; PANTALEO et al. 1993a, b; SCHMITZ et al. 1994; TAMALET et al. 1994).

Bernhard-Nocht Institute for Tropical Medicine, Bernhard-Nocht-Strasse 74, 20359 Hamburg, Germany

The alterations of the follicles as a consequence of persisting GC infection also contribute to the severe immune defect seen in HIV disease. The major immunologic abnormality is the CD4⁺ lymphocyte depletion, but the complex and slowly progressing impairment of the immune system is not due solely to the selective loss of this T cell subset. Functional impairment of antigen-presenting cells and natural killer cells also occurs. Polyclonal B cell activation and impaired humoral response to immunization are manifest in the majority of infected persons long before the depletion of the CD4⁺ lymphocytes (rewieved by ROSENBERG and FAUCI 1993).

GC play a key role during differentiation of B cell response. They develop transiently in response to T-dependent antigenic challenge and are sites of heavy chain class switching, affinity selection of B cells, and generation of B cell memory (reviewed by MACLENNAN et al. 1990, GRAY et al. 1991, LEANDERSON et al. 1992). These events are critically dependent on antigen trapping and retaining function of follicular dendritic cells (FDC). It has also been suggested that antigen held on the surface of FDC could be involved in the maintenance of T cell memory (GRAY and MATZINGER 1991).

The function of FDC in the immune response and the origin and phenotype of these cells will not be stressed in the present review, because these subjects are fully discussed in other chapters of this volume. We will focus on the pathomorphologic changes and the role of these cells in HIV disease. The observation that FDC handle HIV-1 as antigen shed new light on the pathogenesis of HIV infection. Over the past decade, a large number of data has been accumulated that makes a review of the role of FDC in this disease particularly timely.

2 Persistent Generalized Lymphadenopathy

HIV-1 evokes profound functional and morphologic changes of the immune system. Persistent generalized lymphadenopathy (PGL) often develops in individuals who are infected by HIV-1. The morphologic and immunoarchitectural changes of lymph nodes are complex and involve the follicles, the paracortical or interfollicular T cell regions, and cells of the sinuses. It is beyond the scope of this chapter to recapitulate all of these histologic features. A comprehensive review of this topic has recently been published (KNOWLES and CHADBURN 1992). Because HIV-1 replicates and persists in the GC, the histopathologic changes in this compartment of the lymphoid tissue should be summarized briefly.

The morphologic and immunoarchitectural changes of GC represent a spectrum of alterations ranging from follicular hyperplasia to complete loss of the follicles. Examination of sequential biopsies clearly indicates that the changes in the follicles are progressive over the course of HIV disease and correlate with clinical stages (for review see KNOWLES and CHADBOURN 1992). In exuberant follicular hyperplasia, the number and size of the follicles are markedly increased. They are present not only in the cortex, but also in the paracortical or even in the medullary zone. Not seldom do follicles coalesce, giving rise to large, oddly shaped follicles. The mantle zone is often attenuated or absent.

The zonation of the GC is usually present, but the light zone containing centrocytes and T cells may be narrow. Generally, the dark zone with many centroblasts and tingible body macrophages predominates. Mitotic figures are abundant. Immunostaining with monoclonal antibody Ki-67 reveals a high proliferation rate. Plasma cells, sometimes in substantial numbers, and immunoblasts may be present. Clusters of small lymphocytes or invaginations of the extrafollicular parenchyma may break up the follicles into fragments.

The number of intrafollicular $CD4^+$ T cells is either unchanged or slightly decreased when compared to follicular hyperplasia not attributed to HIV infection. The phenotype of these cells, i.e., $CD45RO^+$ and up to 50% also $CD57^+$, is unchanged. Characteristically, the number of $CD8^+$ lymphocytes also expressing the CD45RO antigen (Janossy et al. 1991; Racz et al. 1992) is markedly increased. They are diffusely scattered throughout the follicles and clustered around the vessels.

The FDC network shows significant modifications. The expression of the low-affinity Fc–Immunoglobulin (Ig)E receptor (CD23) on FDC, normally restricted to FDC of the apical light zone (LIU et al. 1991), is upregulated and expands into the basal light zone (Janossy et al. 1991). Sometimes $CD23^+$ FDC virtually fill the whole GC.

The meshwork of FDC exhibits a variable degree of damage. This has been visualized by electron microscopy (Tenner-Racz et al. 1985, 1989; Racz et al. 1986; Piris et al. 1987, Rademakers et al. 1992) and by immunohistochemistry (reviewed by Racz et al. 1989). The usual interwoven concentric staining pattern of the network is interrupted with irregular focal areas of negative staining, giving the meshwork a moth-eaten appearance (Fig.1).

In more advanced stages of the disease, GC undergo regressive changes which range from involution to total loss of the follicles. In some cases hyperplastic and atrophic follicles are present in the same node. The involuted follicles are small and contain only a few centroblasts and centrocytes, and tingible macrophages are usually absent. The number of cells in proliferation is very low. Small vessels with hyalinized walls may penetrate the follicle, and the mantle zone may be thin, absent, or thickened. The number of $CD4^+$ T cells is markedly decreased, whereas that of $CD8^+$ cells is still elevated. Remnants of shrunken and thickened FDC are detectable in these follicles (Fig. 2).

3 Follicular Dendritic Cells Recognize Human Immunodeficiency Virus as Antigen and Initiate Germinal Center Infection

The profound histologic alterations of the follicles and the presence of HIV-1 are interrelated. The key to the infection of GC lies in the antigen-trapping function of

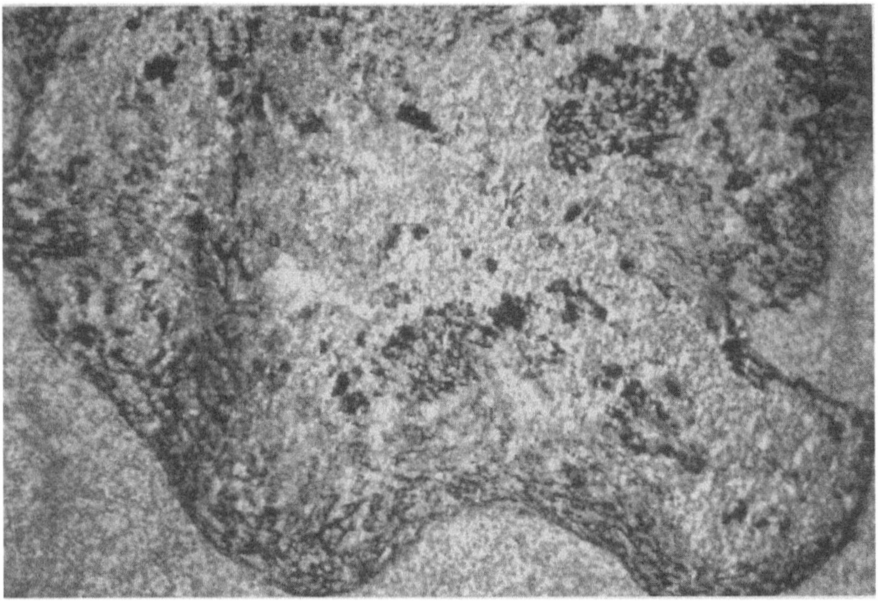

Fig. 1. Partial disruption of follicular dendritic cell (FDC) network in a lymph node with explosive follicular hyperplasia caused by human immunodeficiency virus (HIV)-1. Immunoperoxidase reaction. Magnification, × 90

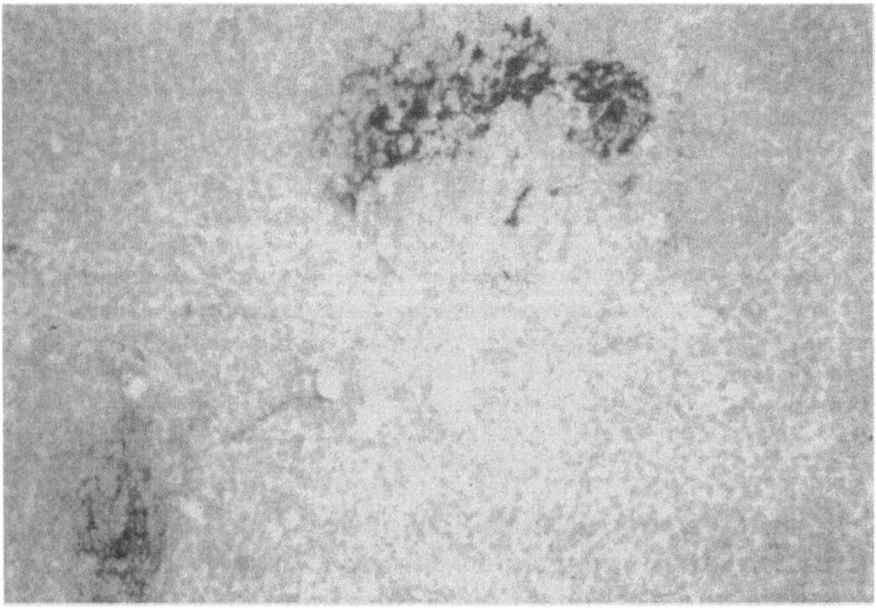

Fig. 2. Remnants of shrunken and thickened follicular dendritic cells (FDC) in an atrophic germinal center. Immunoperoxidase reaction. Magnification, × 112

FDC. The first evidence that FDC of lymph nodes capture HIV-1 was provided by electron microscopic analysis in the mid-1980s (ARMSTRONG and HORNE 1984; ARMSTRONG et al. 1985; TENNER-RACZ et al. 1985, 1986; DIEBOLD et al. 1985; RACZ et al. 1986; CAMERON et al. 1987; PIRIS et al. 1987). These findings were confirmed by immunohistochemical as well as in situ hybridization studies demonstrating excessive and abundant viral proteins and RNA in GC of lymph nodes (TENNER-RACZ et al. 1986, 1988; RACZ et al. 1986; BARONI and UCCINI 1990; BARONI et al. 1986, 1988; BIBERFELD et al. 1986; PARRAVICINI et al. 1986; DIEBOLD et al. 1988a; SCHUURMAN et al. 1988; TENNER-RACZ 1988; PORWIT et al. 1989; EMILIE et al. 1990; FOX et al. 1991; SPIEGEL et al. 1992; PANTALEO et al. 1993a, b; EMBRETSON et al. 1993). Infection of the splenic follicles and the GC of the gut-associated lymphoid tissue has also been documented (DIEBOLD et al. 1988b; LE TOURNEAU et al. 1990; BURKE et al. 1993; RACZ et al. 1993).

Double immunolabeling shows an intimate relationship between HIV proteins, mainly p17 and p24, and FDC (TENNER-RACZ et al. 1986; BARONI et al. 1988; SCHUURMAN et al. 1988; PORWIT et al. 1989). Simultaneous immunostaining with anti-gag and anti-IgM antibodies demonstrates that the staining for both coincides (TENNER-RACZ et al. 1986). Because IgM is bound on and not synthetized by FDC, this finding indicates that HIV-associated antigen is mostly extracellular.

Further evidence for the association of HIV-1 with FDC is that the loss of these cells is accompanied by the loss of HIV antigen, the latter being detectable as long as FDC are still present (TENNER-RACZ et al. 1987; RACZ et al. 1989; PORWIT et al. 1989; BURKE et al. 1993). Even in hyperplastic GC with a partially destroyed FDC network, gag proteins are only present in parts of the follicles that contain FDC (TENNER-RACZ et al. 1986).

The preferential localization of gag proteins and HIV RNA in the light zone of the hyperplastic GC (Figs. 3, 4) also supports the idea of virus trapping by FDC. Previous observations indicate that antigens injected into experimental animals localize in the light zone of the GC (NOSSAL et al. 1966). According to HEINEN and collegues (1984, 1991), one of the hallmarks of FDC is that in vivo captured immune complexes remain fixed on isolated FDC. In agreement with this findings we observed that FDC isolated from infected lymph nodes retained HIV-1 on their surface (Fig. 5) (SCHMITZ et al. 1994; TENNER-RACZ et al. 1994).

A fundamental question is why FDC capture HIV. At this point, it is useful to consider that FDC trap antigen when it is complexed with antibody or able to activate the complement pathway (KLAUS et al. 1980). The captured antigen remains on the cell surface in native form for long periods of time (reviewed by SZAKAL et al. 1989; TEW et al. 1990). Antibody-independent activation of the complement system is triggered by HIV-1 as well as by non-human-associated retroviruses (EBENBLICHER et al. 1991; SÖLDNER et al. 1989; REISINGER et al. 1990; MARSCHANG et al. 1993; THIEBLEMONT et al. 1993; COOPER 1984). The binding of HIV-1 to FDC in vitro is complement dependent and probably involves complement receptors. Anti-HIV-1 antibodies enhance, but do not mediate, HIV-1 binding to FDC (JOLING et al. 1993). FDC carry complement receptors (GERDES and STEIN 1982;

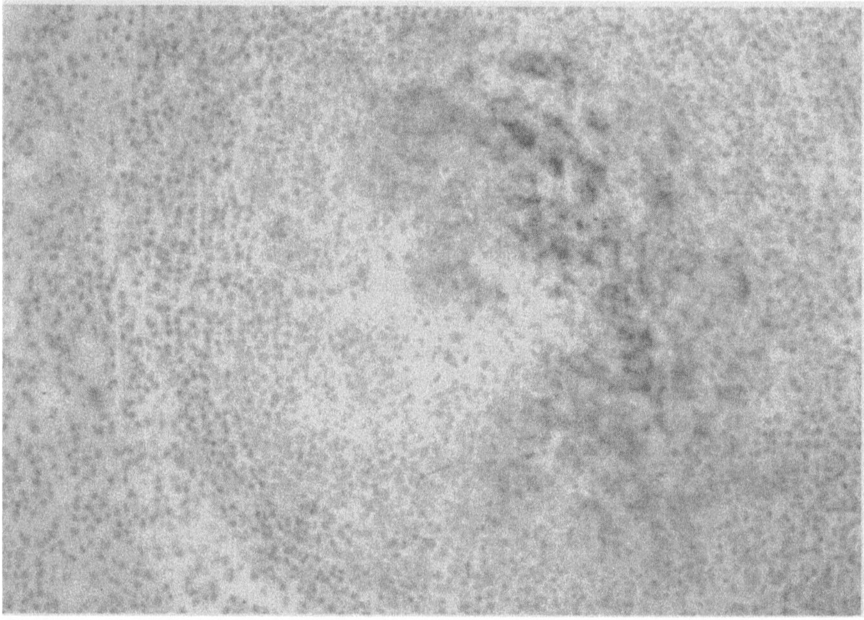

Fig. 3. Distribution of p24 of human immunodeficiency virus (HIV-1) in a follicle. The antigen is localized mainly in the sparsely populated zone of the germinal center. Alkaline phosphatase–antialkaline phosphatase reaction. Magnification, × 112

Fig. 4. Human immunodeficiency virus (HIV)-1 RNA in the sparsely populated zone of a germinal center from a lymph node with explosive follicular hyperplasia. In situ hybridization with an ^{35}S-labeled antisense RNA probe. Epipolarized light. Magnification, × 90

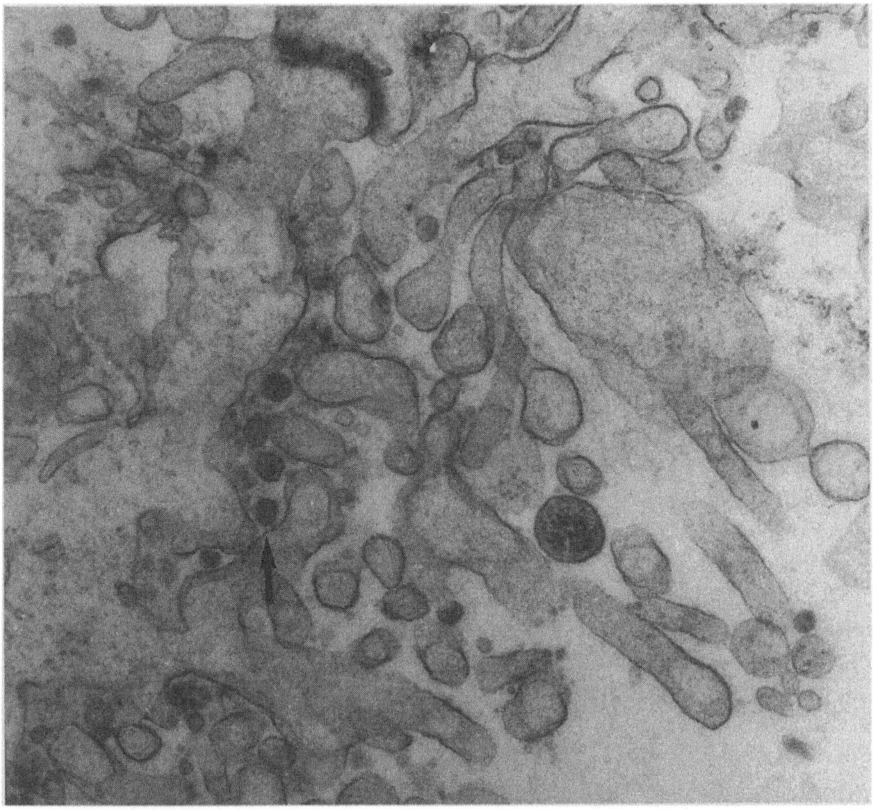

Fig. 5. Human immunodeficiency virus (HIV) particles (*arrow*) on the surface of a follicular dendritic cell (FDC) isolated from a lymph node with explosive follicular hyperplasia. Original magnification, × 20 000

REYNES et al. 1985; SCHRIEVER and NADLER 1992). These findings strongly suggest that the ability of retroviruses to activate the complement pathway is an important, if not the most important, requirement for a virus to localize in the GC.

Indeed, virus trapping by FDC is not restricted to HIV-1 infection as has been reported for several other retroviral infections. Close association of virus particles with FDC was observed in experimental animals infected with Rauscher leukemia virus (HANNA et al. 1970), Abelson leukemia virus (SIEGLER et al. 1973), and feline leukemia virus (TENNER-RACZ et al. 1990). In a murine immunodeficiency model, SZAKAL and coworkers (1992) noted virus trapping of LP-BM5 virus as early as 24 h after i.p. injection.

CHAKRABARTI and collegues (1994) examined early stages of experimental infection with SIV_{mac} in Rhesus macaques and noted the concentration of viral RNA in developing GC between the first and second weeks after infection. The presence of viral particles on the surface of FDC was also demonstrated in cynomolgus monkeys infected with SIV_{sm} (JOLING et al. 1992).

4 Consequence of Virus Trapping by Follicular Dendritic Cells

Two important conclusions follow by necessity if the antigen trapping and retaining function of FDC is responsible for GC infection: (1) the colonization of GC by HIV-1 starts at an early stage of the infection and (2) HIV-1 persists in the GC.

HIV-1 gains access to GC early after infection. Viral proteins and HIV-1 particles can already be seen in acute HIV infection (TENNER-RACZ et al. 1987, 1988b, 1989; TENNER-RACZ 1988; RACZ et al. 1989). Moreover, ARMSTRONG (1991) detected virions in three cases months before seroconversion, indicating that complement-mediated virus trapping starts very early, probably soon after HIV-1 has entered the body. The formation of immune complexes containing HIV after seroconversion does not initiate but rather facilitates virus capture.

Cell-free HIV-1 and infected cells probably reach the lymph nodes via the lymphatics. We were able to demonstrate the presence of p24-positive or HIV RNA-positive cells in the marginal sinus (RACZ 1988; TENNER-RACZ et al. 1988c). In experimental infection of macaque monkeys with SIV_{mac}, an enhanced traffic of productively infected cells through the sinuses was observed before seroconversion (REIMANN et al. 1994; CHAKRABARTI et al. 1994). Viruses delivered into the sinus lumen can reach the follicles, similar to nonviral antigens.

In lymph nodes obtained from patients with acute HIV-1 infection, the

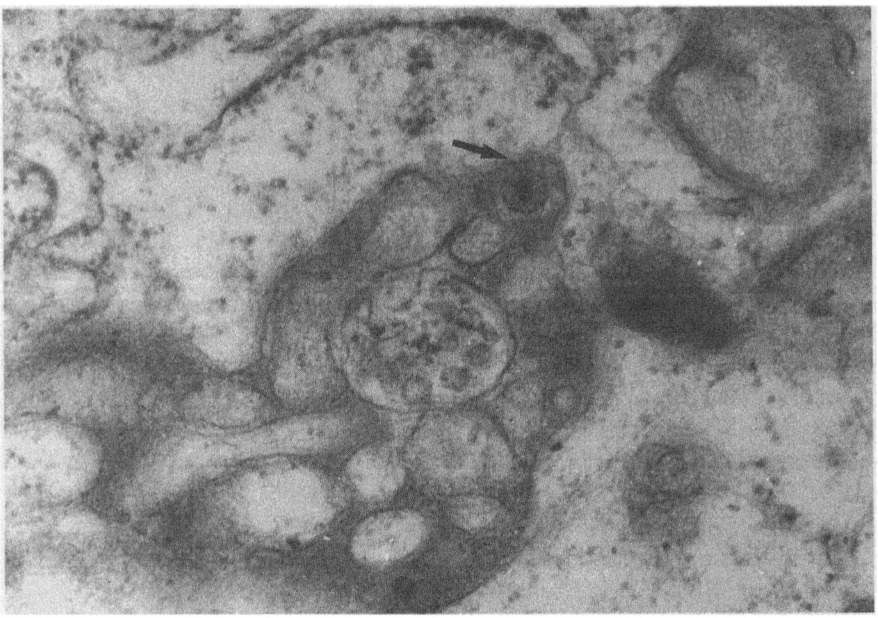

Fig. 6. Lymph node obtained from a patient with acute human immunodeficiency virus (HIV) infection. At this stage only a few virions are present (*arrow*). Original magnificaiton, × 20 000

amount of virions, gag proteins, and viral RNA is very low (Fig. 6). FDC show no ultrastructural or immunohistochemical sign of alteration (TENNER-RACZ et al. 1987, 1989). During the course of the disease, HIV-1 accumulates within the GC. GC of lymph nodes obtained from patients with stage CDC (Centers for Disease Control) III of the disease show many virions (Fig. 7) and large amounts of gag proteins and viral RNA (Figs. 3, 4). Morphologic integrity of the FDC network is not a prerequisite for efficient virus trapping, as was suggested by PANTALEO and coworkers (1993b), because a partial destruction of the meshwork is already present. Investigations performed on repeated lymph node biopsies provided evidence for persisting infection of GC. A partial destruction of the FDC network was present in the first as well as in the second biopsy specimens (TENNER-RACZ et al. 1987, 1988a, 1989).

In addition to diffusely distributed HIV-1 RNA, GC contain cells that replicate the virus. Previous and recent studies demonstrate that cells expressing HIV-1 RNA are mainly located in the follicles (BIBERFELD et al. 1986; TENNER-RACZ 1988; TENNER-RACZ et al. 1988c; SCHUURMAN et al. 1988; EMILIE et al. 1990; FOURNIER et al. 1990; FOX et al. 1991; SPIEGEL et al. 1992; FOX and COTTLER-FOX 1992; PANTALEO et al. 1993b; EMBRETSON et al. 1993). The majority of productively infected cells belong to the $CD4^+$ $CD45R0^+$ $CD57^-$ T cell subset. Up to 10% of these cells per

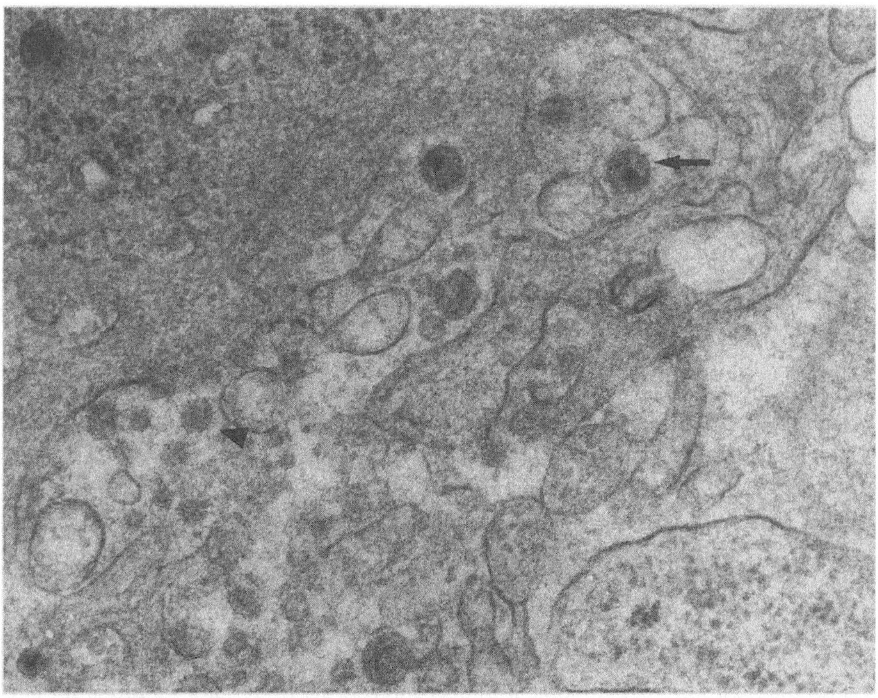

Fig. 7. Many virions (*arrow*) and incomplete particles (*arrowhead*) in the interdendritic extracellular spaces. Stage CDC III of the disease. Original magnification, ×30 000

GC profile replicate the virus (Tenner-Racz et al. 1994). In situ RNA hybridization on serial sections has also demonstrated that all cutting levels contained approximately the same number of labeled cells (Tenner-Racz et al. 1988c). These findings indicate that intrafollicular CD4+ T cells are important virus producers during the phase of clinical latency.

There is a dichotomy between the levels of viral replication in the GC and the extrafollicular parenchyma. Although at the stage of explosive follicular hyperplasia the extrafollicular parenchyma, especially the paracortex, contains a considerable number of latently infected lymphocytes (Embretson et al. 1993), cells expressing viral RNA are scarce at this location. However, in lymph nodes with commencing follicular involution the number of extrafollicular RNA-positive cells usually increases (Tenner-Racz et al. 1993a).

These findings suggest the importance of local immune mechanisms involved in the control of HIV infection of the lymph node. Cytokine production and cytotoxic T cell response play a pivotal role in the modulation of HIV expression. Due to the special function and cellular composition of the GC and the paracortex, the local immunoregulatory network that induces or suppresses HIV gene expression has to be different at these sites. Research efforts in this area are needed to clarify these important aspects of HIV–host interactions.

5 Destruction of Follicular Dendritic Cells

Destruction of FDC network during HIV infection has been reported by many authors (for review see Racz et al. 1989). Although a consistent immunomorphologic finding in PGL, the alteration of FDC is not specific for HIV infection. Disruption of the meshwork has been reported to occur in follicular hyperplasias not attributed to HIV disease (Guettier et al. 1986; Tenner-Racz et al. 1987). Characteristic for HIV-1 infection is, however, that, similar to CD4+ lymphocytes, the loss of FDC is progressive. The ablation of these cells corresponds directly to the fate of GC and therefore has a great impact on the function of the immune system.

The cause of FDC deletion is not known. Two possible mechanisms have been proposed. The first suggests that productive infection and subsequent cell death leads to the loss of FDC. The second postulates a CD8+ T cell-mediated destruction of FDC.

5.1 Infection by Human Immunodeficiency Virus Type 1

FDC can be infected in vitro (Stahmer et al. 1991). In an in situ hybridization study for HIV RNA in combination with immunohistochemistry, Spiegel et al. (1992) found that FDC are productively infected and represent a major reservoir for HIV.

Despite these observations, the majority of data that has accumulated during the last 10 years indicates that although infection of FDC occurs in vivo, its frequency is very low. There have been reports of budding virus on FDC. The authors emphasize, however, that in cases with floride follicular hyperplasia an extensive search was required to find the budding particles (ARMSTRONG and HORNE 1984; CAMERON et al. 1987; TENNER-RACZ et al. 1989; LE TOURNEAU et al. 1990; ARMSTRONG 1991).

PARMENTIER and associates (1990) isolated FDC from a lymph node of a patient with an opportunistic infection and were only able to demonstrate glycoprotein (gp) 41 or HIV RNA in a few cells. JOLING and his coworkers (1993) studied in vitro the binding of HIV to FDC. They found that most of the virions were present extracellularly; gag-immunogold label was only seen in the cytoplasm incidentally. Similarly, proviral DNA analysis by in situ PCR amplification on histologic sections or by nested PCR in limiting dilution assays of purified FDC also showed that FDC are not principal reservoirs of HIV (EMBRETSON et al. 1993; SCHMITZ et al. 1994). Therefore, it is not very likely that the cytopathic effect of HIV-1 substantially contributes to the destruction of FDC.

5.2 Cytotoxic T Cell-Mediated Killing

As mentioned above, one of the most characteristic immunohistologic findings in HIV-1 infection is that $CD8^+$ lymphocytes invade the GC. Almost all of the intrafollicular $CD8^+$ T cells coexpress the CD45R0 isoform of the common leukocyte antigen (RACZ et al. 1992). The focal areas of negative immunostaining within the meshwork of FDC often correspond to areas infiltrated by $CD8^+$ lymphocytes (RACZ et al. 1986; KOOPMAN and PALS 1992). With double immunolabeling, p24-positive cells can occassionally be seen in clusters of $CD8^+$ lymphocytes (TENNER-RACZ et al. 1987). Observations such as these led to postulate the killing of FDC by cytotoxic T cells (CTL; BIBERFELD et al. 1985, 1986; TENNER-RACZ et al. 1987; LAMAN et al. 1989; DEVERGNE et al. 1991)

The function of these intrafollicular $CD8^+$ T cells of memory type is not known. Recent investigations indicate, however, that they belong to the cytotoxic subset. In HIV-1-infected patients the expression of the granzyme B and perforin genes was demonstrated in the intrafollicular $CD8^+$ T cells (DEVERGNE et al. 1991). Another granule-associated protein, TIA-1, which is expressed in CTL and natural killer cells and induces DNA fragmentation (ANDERSON et al. 1990; TIAN et al. 1991; KAWAKAMI et al. 1992), is also present in the intrafollicular $CD8^+$ T Cells (Fig. 8). Nearly the entire $CD8^+$ cell population of the infected GC contains TIA-1 protein. TIA-1-positive cells in mitosis are also seen, indicating that, in addition to invasion, clonal expansion of cytotoxic cells also occurs (TENNER-RACZ et al. 1993).

It is possible that these cells include HIV-specific CTL. Such CTL have been demonstrated in lymph nodes of infected patients (HOFFENBACH et al. 1989; HADIDA et al. 1992). Recently, CHEYNIER and collegues (1994) have shown that HIV-specific

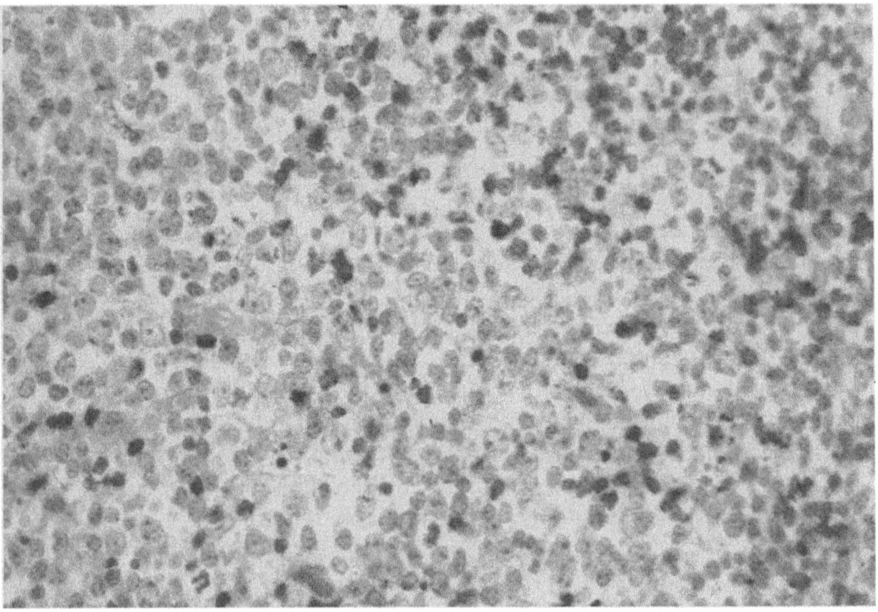

Fig. 8. Cytotoxic cells in a germinal center. Immunostaining with a monoclonal antibody, TIA-1, that recognizes a 15-kDa granule-associated protein whose expression is restricted to cytotoxic T cells and natural killer cells (ANDERSON et al. 1990). Magnification, × 250

CTL infiltrate the white pulp of the spleen. CTL can control the infection of GC and damage cells expressing HIV antigens. Characteristic for CTL function, however, is the recognition of processed antigen which has entered the major histocompatibility complex (MHC) class I antigen presentation pathway. The productively infected FDC can be destroyed by CTL. However, because infected FDC are rare, it is unlikely that a CTL response to processed HIV antigen is responsible for the severe damage and eventual loss of the FDC network. This would only be expected if captured HIV antigen, including fragments of disintegrated virions which should be present in the permanently infected GC, could function as a target for CTL.

It is not unlikely that several factors act in concert to destroy FDC. Beside CTL and the lytic properties of HIV-1, some as yet unidentified factors should be considered. Impaired FDC renewal due to deterioration of the immune system is likely. It has been demonstrated that FDC development in severe combined immunodeficiency mice required transfer of both B and T cells (KAPASI et al. 1993).

The continuous overload of FDC by HIV proteins may also damage mechanisms that physiologically safeguard FDC from membranolysis by captured immune complexes. Recently, it has been demonstrated that FDC express complement-deactivating proteins (LAMPERT et al. 1993; BUTCH et al. 1994). Altered expression of complement-regulatory proteins could exert a deteriorating effect on the integrity of the FDC network.

6 Is Human Immunodeficiency Virus Type 1 Able to Infect Cells Within the Germinal Centers?

The infection of GC has allowed us to postulate that these compartments of the lymphoid tissue represent a major virus reservoir in which infection of trafficking CD4+ T cells occurs. This assumption would be justified if HIV-1 captured by and retained on FDC is infectious.

Although little is known about the virologic events that occur in GC during HIV disease, it can be convincingly argued that infectious virions are present. Because of the migratory capability of lymphocytes, the presence of virus-replicating T cells is no evidence of infection while residing in the GC. These cells may become infected in the GC, or upon entering the follicles latently infected T cells may start to replicate the virus.

Evidence for infectivity of intrafollicular HIV-1 is provided by the observations that, albeit rarely, some FDC replicate the virus. The majority of experimental data indicates that these cells are mesenchymal derived (for review see SCHRIEVER and NADLER 1992). Consequently, FDC represent a nonmigratory and long-lived population. Virus buddings on the FDC surface imply, therefore, that infection of cells occurs within the GC.

Attempts have been undertaken to estimate the viral burden based on the diffuse viral RNA hybridization signals of the GC (FOX et al. 1994). It has to be emphasized that such calculations can be misleading because not only intact, potentially infectious virions are detected by hybridization. Not all captured HIV reaches a target cell. Extracellular particles probably disintegrate. HIV-1 is not known as a stable virus.

More importantly, genetically defective or otherwise noninfectious particles can be generated in or transported to the GC. A large proportion of circulating virus detected by PIATAK et al. (1993) was not culturable. In previous work we also observed morphologically incomplete virions that were composed of amorphous material of varying density and surrounded by envelope in the GC. A well-developed core was not present on serial sections. These particles were present only in association with complete virions and were absent in acute HIV-1 infection (Figs. 6, 7) (TENNER-RACZ et al. 1988b). They can represent defective particles and interfere with target cell infection by nondefective virus, thereby contributing to the slowly progressing disease.

7 Conclusions

During the last decade careful morphologic studies on lymphoid tissues obtained from infected patients have generated important data that opened up a new and surprising vista in the pathogenesis of HIV disease. The biologic strategy of the virus to use the physiologic function of FDC to its benefit places HIV-1 in a highly specialized microenvironment. The antigen trapping and retaining ability of these cells results in virus accumulation. The cellular diversity of GC offers a suitable environment for successful viral transmission, and the rising amounts of HIV-1 increase the incidence of virus-target cell interaction.

One of the consequences of the persisting GC infection and the continuous intrafollicular virus replication could be the generation of HIV variants at this anatomic site. Recent investigations performed on individual microdissected splenic white pulps indicate an exquisite compartmentalization of HIV quasispecies within the lymphoid tissue. The diversity of amino acid sequences within the white pulps was comparable to intra-patient variation (CHEYNIER et al. 1994). This distribution of quasispecies is indicative of local generation. It is highly improbable that FDC of the individual white pulps would discriminate between HIV variants.

Persisting HIV-1 in the follicles also contributes to the severe, complex immunologic disturbance seen in HIV-1 infected individuals. Given the important role that FDC-bound antigens play during a critical stage of B cell differentiation, including generation of B cell memory, the morphologic and functional alterations of the network result in impairment of the humoral immune system. Altered FDC function explains the inability of patients to generate antibodies to new antigens. SZAKAL and collegues (1992) examined the capacity of FDC to trap and retain neoantigens in mice infected with LP-BM5 virus. Virions were present on FDC, but trapping of horseradish peroxidase was significantly reduced within 2 weeks after infection. Antigen captured prior to infection was lost in 4 weeks.

Analyzing the consequences of FDC impairment in the context of HIV-induced immune deficiency, KOOPMAN and PALS (1992) raised a new and rather provocative theory. Because not only B but probably also T cell memory is dependent on unprocessed antigen held on FDC (GRAY and MATZINGER 1991), the authors suggested that FDC loss during HIV-1 infection results in an immune deficiency that selectively affects immunologic memory.

A major challenge for the future is to understand the complex cellular and molecular interactions that occur in HIV-infected lymphoid tissues. It is important to recognize that although experiments on cell suspension provide valuable data, they can not substitute morphologic analysis. The lymphoid architecture represents a specialized microenvironment that directs lymphocyte differentiation and regulates lymphocyte function in vivo. Understanding the relative control of virus replication and the mechanisms that occur while the host mounts an immune response against HIV-1 will provide the basis for more rational therapeutic strategies.

References

Anderson P, Nagler-Anderson C, O'Brien C, Levine H, Watkons S, Slayter HS, Blue ML, Schlossman SF (1990) A monoclonal antibody reactive with a 15 kDa cytoplasmic granule associated protein defines a subpopulation of CD8+ lymphocytes. J Immunol 144: 574–582

Armstrong JA (1991) Ultrastructure and significance of the lymphoid tissue lesions in HIV infection. In: Racz P, Dijkstra CD, Gluckman JC (eds) Accessory cells in HIV and other retroviral infections. Karger, Basel, pp 69–82

Armstrong JA, Horne R (1984) Follicular dendritic cells and viruslike particles in AIDS-related lymphadenopathy. Lancet ii: 370–372

Armstrong JA, Dawkins RL, Horne R (1985) Retroviral infection of accessory cells and the immunological paradox in AIDS. Immunol Today 6: 121–122

Baroni CD, Uccini S (1990) HIV and EBV expression in lymph nodes: immunohistology and in situ hybridization. In: Racz P, Hasse AT, Gluckman JC (eds) Modern pathology of AIDS and other retroviral infections. Karger, Basel, pp 99–109

Baroni CD, Pezzella F, Mirolo M, Ruco LP, Rossi GB (1986) Immunohistochemical demonstration of p24 HTLV-III major core protein in different cell types within lymph nodes from patients with lymphadenopathy syndrome (LAS). Histopathology 10: 5–13

Baroni CD, Pezzella F, Pezzella M, Macchi B, Vitolo D, Uccini S, Ruco LP (1988) Expression of HIV in lymph node cells of LAS patients. Immunohistology, in situ hybridization and identification of target cells. Am J Pathol 133: 498–506

Biberfeld P, Porwit-Ksiazek A, Böttiger B, Morfeldt-Mansson L, Biberfeld G (1985) Immunohistopathology of lymph nodes in HTLV-III infected homosexuals with persistent adenopathy or AIDS. Cancer Res [Suppl] 45: 465s–470s

Biberfeld P, Chayt KJ, Marselle LM, Biberfeld G, Gallo RC, Harper M (1986) HTLV-III expression in infected lymph nodes and relevance to pathogenesis of lymphadenopathy. Am J Pathol 125: 436–442

Burke AP, Benson W, Ribas JL, Anderson D, Chu WS, Smialek J, Virmani R (1993) Postmortem localization of HIV-1 RNA by in situ hybridization in lymphoid tissues of intravenous drug addicts who died unexpectedly. Am J Pathol 142: 1701–1713

Butch AW, Hug BA, Nahm MH (1994) Properties of human follicular dendritic cells purified with HJ2, a new monoclonal antibody. Cell Immunol 155: 27–41

Cameron PU, Dawkins RL, Armstrong JA, Bonifacio E (1987) Western blot profiles, lymph node ultrastructure, and viral expression in HIV-infected patients. A correlative study. Clin Exp Immunol 68: 465–478

Chakrabarti L, Isola P, Cumont MC, Clasessens-Maire MA, Hurtrel M, Montagnier L, Hurtrel B (1994) Early stages of simian immunodeficiency virus infection in lymph nodes. Evidence for high viral load and successive populations of target cells. Am J Pathol 144: 1226–1237

Cheynier E, Henrichwark S, Hadida F, Pelletier E, Oksenhendler E, Autran B, Wain-Hobson S (1994) HIV and T cell expansion in splenic white pulps is accompanied by infiltration of HIV-specific cytotoxic T lymphocytes. Cell 78: 373–387

Cooper NR (1984) The role of the complement system in host defense against virus diseases. In: Notkins AL, Oldstone MBA (eds) Concepts in viral pathogenesis. Springer, Berlin Heidelberg New York, pp 20–25

Devergne O, Peuchmaur M, Crevon MC, Trapani JA, Maillot MC, Galanaud P, Emilie D (1991) Activation of cytotoxic cells in hyperplastic lymph nodes from HIV-infected patients. AIDS 5: 1071–1079

Diebold J, Marche CL, Audouin J, Aubert JP, Le Tourneau A, Bouton CL, Reynes M, Wizniak J, Capron F, Tricottet V (1985) Lymph node modification in patients with the acquired immunodeficiency syndrome (AIDS) or with AIDS-related complex (ARC). A histological, immunohistological and ultrastructural study of 45 cases. pathol Res Pract 180: 590–611

Diebold J, Audouin J, Le Tourneau A (1988a) Lymph node biopsy in the diagnosis of persistent generalized lymphadenopathy syndrome in patients at risk for acquired immunodeficiency. Prog Surg Pathol 8: 55–60

Diebold J, Audouin J, Le Tourneau A (1988b) Lymphoid tissue changes in HIV-infected patients. Lymphology 21: 22–27

Ebenblicher CF, Thielens NM, Vornhagen R, Marschang P, Arlaud GJ, Dierich MP (1991) Human Immuno-deficiency virus type-1 activates the classical pathway of complement by direct C1 binding through specific site in transmembrane glycoprotein gp41. J Exp Med 174: 1417–1424

Embretson J, Zupancic M, Ribas JL, Burke A, Racz P, Tenner-Racz K, Haase AT (1993) Massive covert infection of helper T lymphocytes and macrophages by human immunodeficiency virus during the incubation period of AIDS. Nature 262: 359–362
Emilie D, Peuchmaur M, Maillot MC, Crevon MC, Brousse N, Delfraissy JF, Dormont J, Galanaud P (1990) Production of interleukins in human immunodeficiency virus-1-replicating lymph nodes. J Clin Invest 86: 148–159
Fournier JG, Prevot S, Audouin J, Le Tourneau A, Lebon P, Diebold J (1990) Fine analysis of HIV-1-RNA detection by in situ hybridization. In: Racz P, Haase AT, Gluckman JC (eds) Modern pathology of AIDS and other retroviral infections. Karger, Basel, pp 62–68
Fox CH, Cottler-Fox M (1992) The pathobiology of HIV infection. Immunol Today 13: 353–356
Fox CH, Tenner-Racz K, Racz P, Fripo A, Pizzo PA, Fauci AS (1991) Lymphoid germinal centers are reservoirs of human immunodeficiency virus type 1 RNA. J Infect Dis 164: 1051–1057
Fox CH, Hoover S, Currall VR, Bahre HJ, Cottler-Fox M (1994) HIV in infected lymph nodes. Nature 370: 256
Gerdes J, Stein H (1982) Complement (C3) receptors on dendritic reticulum cells of normal and malignant lymphoid tissue. Clin Exp Immunol 48: 348–352
Gray D, Matzinger P (1991) T cell memory is short-lived in the absence of antigen. J Exp Med 174: 969–974
Gray D, Kosco M, Stockinger B (1991) Novel pathway of antigen presentation for the maintenance of memory. Int Immunol 3: 141
Guettier C, Gatter KC, Heryet A, Mason DY (1986) Dendritic reticulum cells in reactive lymph nodes and tonsils: an immunohistochemical study. Histopathology 10: 15–24
Hadida F, Parrot A, Kieny MP, Sadat-Sowdi B, Mayaud C, Debrè P, Autran B (1992) Carboxy-terminal and central regions of human immunodeficiency virus-1 nef recognized by cytotoxic T lymphocytes from lymphoid organs. J Clin Invest 89: 53–60
Hanna MG, Szakal AK, Tyndall RL (1970) Histoproliferative effect of Rauscher leukemia virus on lymphatic tissue: histological and ultrastructural studies of germinal centers and their relation to leukemogenesis. Cancer Res Arch 30: 1748–1763
Heinen E, Lilet-Leclercq C, Mason DY, Stein H, Boniver J, Radoux D, Kinet-Denoel C, Simar LJ (1984) Isolation of follicular dendritic cells from human tonsils or adenoids. II. Immunocytochemical characterization. Eur J Immunol 14: 267–273
Heinen E, Cormann N, Tsunoda R, Kinet-Denoel C, Simar LJ (1991) Ultrastructural and functional aspects of follicular dendritic cells in vitro. In: Racz P, Dijkstra CD, Gluckman JC (eds) Accessory cells in HIV and other retroviral infections. Karger, Basel, pp 1–8
Hoffenbach A, Langlade-Demoyen P, Dadaglio G, Vilmer E, Michel F, Mayaud C, Autran B, Plata F (1989) Unusually high frequencies of HIV-specific cytotoxic T lymphocytes in humans. J Immunol 142: 452–462
Janossy G, Bofill M, Johnson M, Racz P (1991) Changes of germinal center organization in HIV-positive lymph nodes. In: Racz P, Dijsktra C, Gluckman JC (eds) Accessory cells in HIV and other retroviral infections. Basel, Karger, pp 111–123
Joling P, van Wichen DF, Parmentier HK, Biberfeld P, Böttiger D, Tschopp J, Rademakers LH, Schuurman HJ (1992) Simian immunodeficiency virus (SIV_{sm}) infection of cynomolgus monkeys: effect on follicular dendritic cells in lymphoid tissue. AIDS Res Hum Retroviruses 8: 2021–2030
Joling P, Bakker LJ, van Strijp JAG, Meerloo T, de Graaf L, Dekker MEM, Goudsmit J, Verhoef J, Schuurman HJ (1993) Binding of human immunodeficiency virus type-1 to follicular dendritic cells in vitro is complement dependent. J Immunol 150: 1065–1073
Kapasi ZF, Burton GF, Schultz LD, Tew JG, Szakal AK (1993) Induction of functional follicular dendritic cell development in severe combined immunodeficiency mice. Influence of B and T cells. J Immunol 150: 2648–2658
Kawakami A, Tian Q, Duan X, Streuli M, Schlossman SF, Anderson P (1992) Identification and functional characterization of a TIA-1-related nucleolysis. Proc Natl Acad Sci USA 89: 8681–8685
Klaus GGB, Humphrey JH, Kunkl A, Dongworth DW (1980) The follicular dendritic cell: its role in antigen presentation in the generation of immunological memory. Immunol Rev 53: 3–28
Knowles DM, Chadburn A (1992) Lymphadenopathy and lymphoid neoplasms associated with the acquired immune deficiency syndrome (AIDS). In: Knowles DM (ed) Neoplastic hematopathology. Williams and Wilkins, Baltimore, pp 773–783
Koopman G, Pals ST (1992) Cellular interactions in the germinal center: role of adhesion receptors and significance for the pathogenesis of AIDS and malignant lymphoma. Immunol Rev 126: 21–45
Laman JD, Classen E, van Rooijen N, Boersma WJA (1989) Immune complexes on follicular dendritic cells as a target for cytolytic cells in AIDS. AIDS 3: 543–544

Lampert IA, Schofield JB, Amlot P, van Noorden S (1993) Protection of germinal centres from complement attack: decay-accelerating factor (DAF) is a constitutive protein on follicular dendritic cells. A study in reactive and neoplastic follicles. J Pathol 170: 115–120

Le Tourneau A, Prevot S, Diebold J, Audouin J, Carnot F, Audrouin C, Tourani JM, Bouchard I, Espinoza P, Kazatchkine M, Andrieu JM (1990) Electron microscopy study of the spleen in HIV-infected patients. In: Racz P, Dijkstra CD, Gluckman JC (eds) Accessory cells in HIV and other retroviral infections. Karger, Basel, pp 98–110

Leanderson T, Källberg E, Gray D (1992) Expansion, selection and mutation of antigen-specific B cells in germinal centers. Immunol Rev 126: 47–61

Liu YJ, Cairns JA, Holder MJ, Abbot S, Jansen KU, Bonnefoy JY, Gordon J, MacLennan ICM (1991) Recombinant 25 KD CD23 and interleukin 1 promote survival of germinal centre B cells: evidence for bifurcation in the development of centrocytes rescused from apoptosis. Eur J Immunol 21: 1107–1114

MacLennan ICM, Liu YJ, Oldsfeld S, Zhang J, Lane PJL (1990) The evolution of B-cell clones. In: Gray D, Sprent J (eds) Immunological memory. Springer, Berlin Heidelberg New York, pp 39–63 (Current topics in microbiology and immunology, vol 159)

Marschang P, Gürtler L, Tötsch M, Thielens NM, Arlaud GJ, Hittmair A, Katinger H, Dierich MP (1993) HIV-1 and HIV-2 isolates differ in their ability to activate the complement system on the surface of infected cells. AIDS 7: 903–910

Nossal GJV, Austin CM, Pye J, Mitchell J (1966) Antigens in immunity. XII. Antigen trapping in the spleen. Int Arch Allergy 29: 368–383

Pantaleo G, Graziosi C, Fauci AS (1993a) The immunopathogenesis of human immunodeficiency virus infection. N Engl J Med 328: 327–335

Pantaleo G, Graziosi C, Demarest JF, Butini L, Montroni M, Fox CH, Orenstein JM, Kotler DP, Fauci AS (1993b) HIV infection is active and progressive in lymphoid tissue during the clinically latent stage of disease. Nature 362: 355–358

Parmentier HK, van Wichen DF, Sie-Go DMDS, Goudsmit J, Borleffs JJC, Schuurman HJ (1990) Infection and virus production in follicular dendritic cells in lymph nodes. A case report with analysis of isolated follicular dendritic cells. Am J Pathol 137: 247–251

Parravicini CL, Yago L, Costanzi GC (1986) Follicle lysis in lymph node biopsies from patients with persistent generalized lymphadenopathy. Blood 8: 576–577

Piatak M Jr, Saag MS, Yang LC, Clark SJ, Kappes JC, Luk KC, Hahn BH, Shaw GM, Lifson JD (1993) High levels of HIV-1 in plasma during all stages of infection determined by competitive PCR. Science 259: 1749–1754

Piris MA, Rivas C, Morente M, Rubio C, Martin C, Olivia H (1987) Persistent and generalized lymphadenopathy: a lesion of follicular dendritic cells? An immunohistologic and ultrastructural study. Am J Clin Pathol 87: 716–724

Porwit A, Böttiger B, Pallesen G, Bodner A, Biberfeld P (1989) Follicular involution in HIV lymphadenopathy. A morphometric study. Acta Pathol Microbiol Scand 7: 153–165

Racz P (1988) Molecular, biologic, immunohistochemical and ultrastructural aspects of lymphatic spread of the human immunodeficiency virus. Lymphology 21: 28–35

Racz P, Tenner-Racz K, Kahl C, Feller AC, Kern P, Dietrich M (1986) The spectrum of morphologic changes in lymph nodes from patients with AIDS or AIDS-related complex. Prog Allergy 37: 81–181

Racz P, Tenner-Racz K, Schmidt H (1989) Follicular dendritic cells in HIV-induced lymphadenopathy and AIDS. Acta Pathol Microbiol Scand 97 [Suppl 8]: 16–23

Racz P, Tenner-Racz K, van Vloten F, Schmidt H, Dietrich M, Gluckman JC, Letvin NL, Janossy G (1990) Lymphatic tissue changes in AIDS and other retrovirus infections: tools and insights. Lymphology 23: 85–91

Racz P, Tenner-Racz K, van Vloten F, Letvin NL, Janossy G (1992) $CD8^+$ lymphocytes response in lymph nodes from patients with HIV infection. In: Racz P, Letvin NL, Gluckman JC (eds) Cytotoxic T lymphocytes in HIV and other retroviral infections. Karger, Basel, pp 162–173

Racz P, Moncada LA, Schmidt H, von Stemm A, Hartmann MG, Dietrich M, Tenner-Racz K (1993) Lymphatic follicles of rectal mucosa harbor HIV-1. Lab Invest 68: 106A

Rademakers IHPM, Schuurman HJ, de Frankrijken JF, van Ooyen A (1992) Cellular composition of germinal centers in lymph nodes after HIV-1 infection: evidence for an inadequate support of germinal center B lymphocytes by follicular dendritic cells. Clin Immunol Immunopathol 62: 148–159

Reimann KA, Tenner-Racz K, Racz P, Montefiori DC, Yasutomi Y, Lin W, Ransil BJ, Letvin NL (1994) Immunopathogenic events in the acute infection of Rhesus monkeys with simian immunodeficiency virus of macaques. J Virol 68: 2362–2370

Reisinger EC, Vogetseder W, Berzow D, Köfler D, Bitterlich G, Lehr HA, Wachter H, Dierich MP (1990) Complement-mediated enhancement of HIV-1 infection of the monoblastoid cell U937. AIDS 4: 961–965

Reynes M, Aubert JP, Cohen JHM, Audouin J, Tricottet V, Diebold J, Kazatchkine MD (1985) Human follicular dendritic cells express CR1, CR2, and CR3 complement receptor antigens. J Immunol 135: 2687–2694

Rosenberg ZF, Fauci AJ (1993) Immunology of HIV infection. In: Paul WE (ed) Fundamental immunology, 3rd edn. Raven, New York, pp 1375–1397

Schmitz J, van Lunzen J, Tenner-Racz K, Großschupff G, Racz P, Schmitz H, Dietrich M, Hufert FT (1994) Follicular dendritic cells retain HIV-1 particles on their plasma membrane, but are not productively infected in asymptomatic patients with follicular hyperplasia. J Immunol 153: 1352–1359

Schnittman SM, Greenhouse JJ, Lane HC, Pierce PF, Fauci AS (1991) Frequent detection of HIV-1 specific mRNAs in infected individuals suggests ongoing active viral expression in all stages of disease. AIDS Res Human Retroviruses 7: 361–367

Schriever F, Nadler LM (1992) The central role of follicular dendritic cells in lymphoid tissue. Adv Immunol 51: 243–284

Schuurman HJ, Krone WJA, Broekhuizen R, Goudsmit J (1988) Expression of RNA and antigens of human immunodeficiency virus type-1 (HIV-1) in lymph nodes from HIV-1 infected individuals. Am J Pathol 133: 516–524

Siegler R, Lane I, Frisch Y, Moran S (1973) Early response of lymph node cells to Abelson leukemia virus. Lab Invest 29: 273–277

Sölder BM, Schultz TF, Hengster P, Löwer J, Lacher C, Bitterlich G, Kurth R, Wachter H, Dierich MP (1989) HIV and HIV-1 infected cells differentially activate the human complement system independent of antibody. Immunol Lett 22: 135–146

Spiegel H, Herbst H, Niedobitek G, Foss HD, Stein H (1992) Follicular dendritic cells are a major reservoir for human immunodeficiency virus type 1 in lymphoid tissues facilitating infection of CD4+ T-helper cells. Am J Pathol 140: 15–22

Stahmer I, Zimmer JP, Ernst M, Fenner T, Finnern R, Schmitz H, Flad HD, Gerdes H (1991) Isolation of normal human follicular dendritic cells and CD4-independent in vitro infection by human immuno-deficiency virus (HIV-1). Eur J Immunol 21: 1873–1878

Szakal AK, Kosco MH, Tew JG (1989) Microanatomy of lymphoid tissue during the induction and maintenance of humoral immune response: structure function relationship. Annu Rev Immunol 7: 91–109

Szakal AK, Kapasi ZF, Masuda A, Tew JG (1992) Follicular dendritic cells in the alternative antigen transport pathway: microenvironment, cellular events, age and retrovirus related alterations. Semin Immunol 4: 257–265

Tamalet C, Lafeuilade A, Yahi N, Vignoli C, Tourres C, Pellegrino P, de Micco P (1994) Comparison of viral burden and phenotype of HIV-1 isolates from lymph nodes and blood. AIDS 8: 1083–1088

Tenner-Racz K (1988) Human immunodeficiency virus-associated changes in germinal centers of lymph nodes and relevance to impaired B-cell function. Lymphology 21:36–43

Tenner-Racz K, Racz P, Dietrich M, Kern P (1985) Altered follicular dendritic cells and virus-like particles in AIDS and AIDS-related lymphadenopathy. Lancet i: 105–106

Tenner-Racz K, Racz P, Bofill M, Schulz-Meyer A, Dietrich M, Kern P, Weber J, Pinching AJ, Veronese-Dimarzo F, Popovic M, Klatzmann D, Gluckman JC, Janossy G (1986) HTLV-III/LAV viral antigens in lymph nodes of homosexual men with persistent generalized lymphadenopathy and AIDS. Am J Pathol 123: 9–15

Tenner-Racz K, Racz P, Dietrich M, Kern P, Janossy G, Veronese-Dimarzo F, Klatzmann D, Gluckman JC, Popovic M (1987) Monoclonal antibodies to human immunodeficiency virus: their relation to the patterns of lymph node changes in persistent generalized lymphadenopathy and AIDS. AIDS 1: 95–104

Tenner-Racz K, Racz P, Gluckman JC, Popovic M (1988a) Cell-free HIV in lymph nodes of patients with AIDS and generalized lymphadenopathy. N Engl J Med 318: 49–50

Tenner-Racz K, Racz P, Gartner S, Dietrich M, Popovic M (1988b) Atypical virus particles in HIV-1-associated persistent generalised lymphadenopathy. Lancet i: 774–775

Tenner-Racz K, Racz P, Schmidt H, Dietrich M, Kern P, Louie A, Gartner S, Popovic M (1988c) Immunohistochemical, electron microscopic and in situ hybridization evidence for the involvement of lymphatics in the spread of HIV-1. AIDS 2: 299–309

Tenner-Racz K, Racz P, Gartner, S, Ramsauer J, Dietrich M, Gluckman JC, Popovic M (1989) Ultrastructural analysis of germinal centers in lymph nodes of patients with HIV-1-induced persistent generalized lyhmphadenopathy: evidence for persistence of infection. In: Rotterdam H, Sommers SC, Racz P, Mayer PR (eds) Progress in AIDS pathology, vol 1. Field and Wood, New York, pp 29–40

Tenner-Racz K, Racz P, Taveres LM, H Schmidt, Noronha F (1990) Ultrastructural aspects of lymphadenopathy induced by the feline leukemia virus: evidence for intrafollicular virus replication and spread of infection through the lymphatics. In: Racz P, Haase AT, Gluckman JC (eds) Modern pathology of AIDS and other retroviral infections. Karger, Basel, pp 161–176

Tenner-Racz K, von Stemm A, Schmidt H, Schoenberger OL, Racz P (1993a) High percentage of lymphocytes are productively infected in the germinal centers. Lab Invest 68: 107A

Tenner-Racz K, Racz P, Thomé C, Meyer CG, Anderson PJ, Schlossman SJ, Letvin NL (1993b) Cytotoxic effector cell granules recognized by the monoclonal antibody TIA-1 are present in $CD8^+$ lymphocytes in lymph nodes of HIV-1-infected patients. Am J Pathol 142: 1750–1758

Tenner-Racz K, von Stemm AMR, Gühlk B, Schmitz J, Racz P (1994) Are follicular dendritic cells, macrophages and interdigitating cells of the lymphoid tissue productively infected bt HIV? Res Virol 145: 177–182

Tew JG, Kosco MH, Burton GF, Szakal AK (1990) Follicular dendritic cells as accessory cells. Immunol Rev 117: 185–211

Thieblemont N, Haffner-Cavaillon N, Weiss L, Maillet F, Kazatchkine MD (1993) Complement activation by gp160 glycoprotein of HIV-1. AIDS Res Human Retroviruses 9: 229–233

Tian Q, Streuli M, Saito H, Schlossman SF, Anderson P (1991) A polyadenylate binding protein localized to the granules of cytolytic lymphocytes induces DNA fragmentation in target cells. Cell 67: 629–639

Follicular Dendritic Cells and Infection by Human Immunodeficiency Virus Type 1— A Crucial Target Cell and Virus Reservoir

H.-J. Schuurman[1,2], P. Joling[2], D.F. van Wichen[2], L.H.P.M. Rademakers[2], R. Broekhuizen[2], R.A. de Weger[2], J.G. van den Tweel[2], and J. Goudsmit[3]

1 Introduction: Conventional Histology of Lymph Nodes After Human Immunodeficiency Virus Type 1 Infection

Even before human immunodeficiency virus type 1 (HIV-1) was recognized as the causative agent of the acquired immunodeficiency syndrome (AIDS), generalized persistent lymphadenopathy was considered as one of the symptoms in this condition. The atypical histologic features of the lymph nodes, mainly the hyperplasia of follicles, were also identified at that time. After the discovery of HIV-1 and the introduction of serology to assess the status of infection, clinicopathologic evaluations could be performed. Schuurman et al. (1985) differentiated three main stages in lymph node abnormalities, which were similar to stages described by Baroni and Uccini (1990), Biberfeld et al. (1985), Burns et al. (1985), Garcia et al. (1986), Janossy et al. (1985), Öst et al. (1989), Porwit et al. (1989), Rácz et al. (1986, 1990), and Wood (1990) (Fig. 1):

1. Persistent generalized lymphadenopathy develops in the first phase after infection, either as the only symptom or in combination with constitutional symptoms such as persistent fever, weight loss, night-sweat, and diarrhea.

[1]Preclinical Research, Sandoz Pharma Ltd., Basel, Switzerland
[2]Division of Histochemistry and Electron Microscopy, Department of Pathology, University Hospital Utrecht, The Netherlands
[3]Human Retroviral Laboratory, Academic Medical Center, Amsterdam, The Netherlands

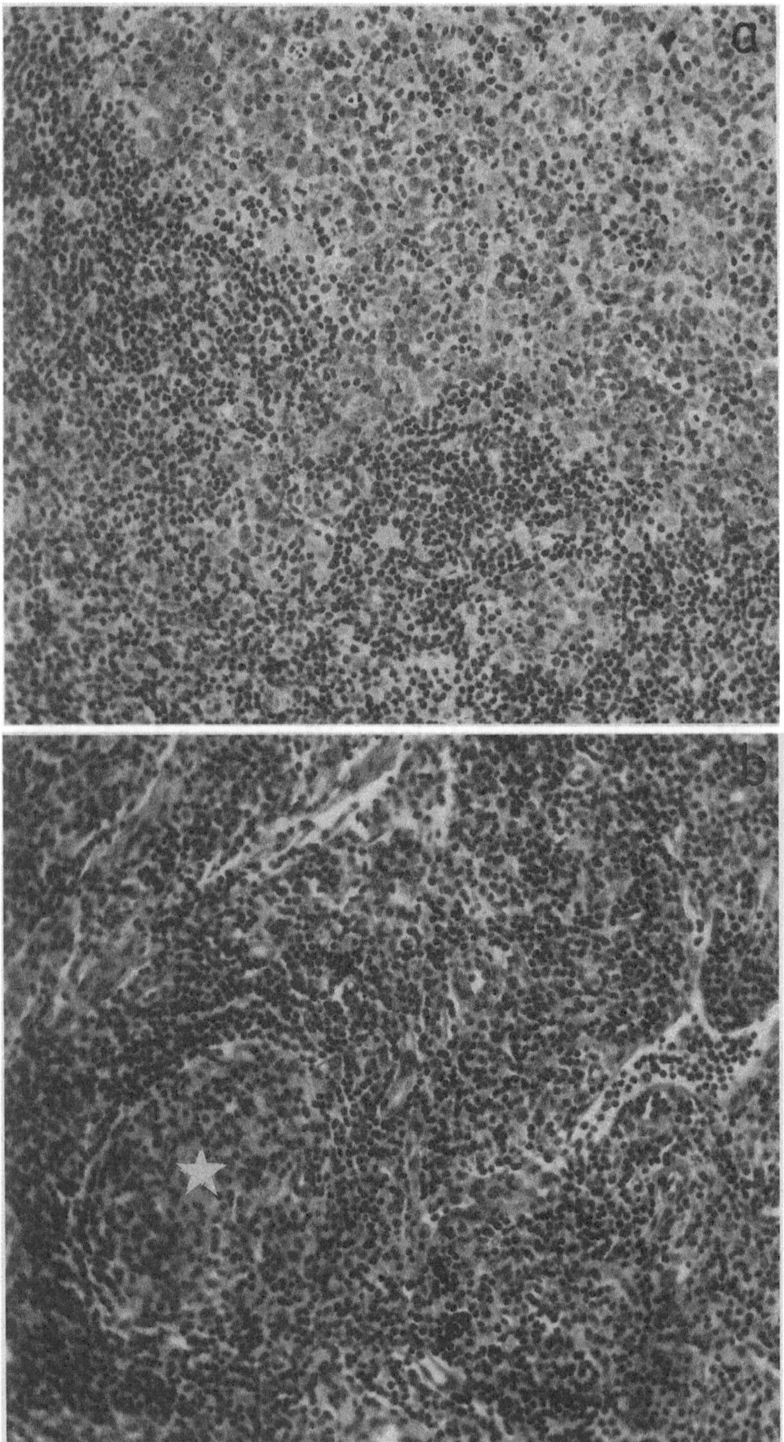
a
b

Histology of the swollen lymph nodes shows hyperplasia of the follicles. The follicles not only are large with pronounced germinal centers, but also often show a bizarre shape of germinal centers with indentations and fragmentations of the follicular dendritic cell (FDC) meshwork. These indentations and fragmentations are visible in the histologic section, as these show the accumulation of small-sized lymphocytes, and not the pale-staining large germinal center cells (Fig. 1a). The mantle of the follicles is of variable size, and individual follicles can vary from almost absent to quite large. The interfollicular areas in the swollen nodes do not manifest major abnormalities, but can manifest vascular endothelial cell proliferation like that seen in angioimmunoblastic lymphadenopathy. According to the disease classification by the CENTERS FOR DISEASE CONTROL (CDC; 1986), follicle hyperplasia occurs in either CDC group III (persistent generalized lymphadenopathy) or CDC IVA.

2. In the subsequent stage, lymphadenopathy is less pronounced. In histology, follicles show a larger extent of fragmentation and degeneration. Some resemblance with angioimmunoblastic lymphadenopathy can be more evident during the process of follicle degeneration, as visualized by vascular proliferation, lymphocyte depletion, and emergence of blastoid cells (Fig. 1b). In CDC subgroups of CDC IV, when patients have neurologic disease, opportunistic infections, or neoplasia associated with the diagnosis AIDS, lymph nodes show either follicular hyperplasia or follicular degeneration.
3. In the last stage, lymph nodes become very small. Histologically, there is follicle atrophy with hyalinized remnants of the original architecture. The node is often depleted of lymphocytes in this terminal stage; plasma cells form one of the main lymphocyte subsets still present in this stage. Lymph node atrophy is mainly seen in the terminal phase of AIDS.

According to these histologic features, the follicle is the main compartment in the lymph node in which changes occur. This phenomenon is unexpected in the view of the fact that T lymphocytes and macrophages are the first target cells for infection and that the follicle is considered a typical B lymphocyte environment. Only after a more in depth analysis of lymph nodes, using advanced histochemical procedures and electron microscopy, and studies on isolated lymph node stromal cells in suspension did it become evident that follicles actually play a crucial role in accumulating viral components. This review presents a survey of these studies in relation to the "normal" hyperplastic lymph node follicle.

Fig. 1 a,b. Histology of lymph nodes from patients after human immunodeficiencny virus (HIV)-1 infection. **a** Follicular hyperplasia showing part of a large-sized follicle (*upper right*) with pronounced germinal center and mantle (*bottom left*). Note the irregular shape of the germinal center, with indentation. x 95. **b** Follicular degeneration, showing a burned-out appearance of the germinal center (*asterisk*), with hyalinized material. Note the vascular proliferation in the interfollicular area. x 95. Formalin-fixed material, embedded in paraffin, hematoxylin-eosin staining

2 Histophysiologic Features of Follicles in the Normal Human Lymph Node and Tonsil

The structural components of follicles are FDC. The view that the origin of these accessory cells is mesenchymal (HEUSERMANN et al. 1980; HUMPHREY et al. 1984; IMAZEKI et al. 1992; KLAUS et al. 1980) is supported by their immunoreactivity for vimentin and desmin, these being mesenchymal markers (RADEMAKERS et al. 1992b). Their emergence in tissue normally requires the presence of B lymphocytes (YOSHIDA et al. 1994). The main function of FDC is antigen presentation to B lymphocytes in the follicle. Reviews on this aspect have been published by JANOSSY et al. (1991), KOOPMAN and PALS (1992), KOSCO and GRAY (1992), MACLENNAN et al. (1992), SCHRIEVER and NADLER (1992), SZAKAL et al. (1989, 1992), and WACKER et al. (1990). This antigen presentation is not controlled by major histocompatibility complex (MHC) molecules. B lymphocytes recognize antigen in nominal form entrapped in the labyrinth of protrusions of FDC in the germinal center, either solitary or in the form of immune complexes, and subsequently differentiate with the help of ($CD4^+$, helper phenotype) T cells or their products. The continuous exposure of antigen to B lymphocytes in the germinal center has been associated with immunologic memory at the B cell level: after the disappearance of the antigen from the germinal center, B cell memory also fades.

Studies on FDC in suspension have been hampered by the complications in isolating the cells from tissue and the absence of FDC lines. Isolation procedures that have been developed include the disintegration of the stroma, harvesting of either whole tissue or follicles by a stereomicroscope, and the use of enzymatic digestion with collagenase and DNAse (CLARK et al. 1992; ENNAS et al. 1989; LILET-LECLERCQ et al. 1984; LINDHOUT et al. 1993; PARMENTIER et al. 1991a; PETRASCH et al. 1990; SCHNIZLEIN et al. 1985; SCHMITZ et al. 1993; SCHRIEVER et al. 1989; SELLHEYER et al. 1989). Subsequently, sedimentation over a discontinuous bovine serum albumin (BSA) gradient and Percoll density gradient is performed. In this way, suspensions are obtained that are rich in FDC, identified in cytological preparations as large, binucleated cells with dendritic processes (JOLING et al. 1993; KOOPMAN et al. 1991; PARMENTIER et al. 1991a). These suspensions are contaminated mainly by B cells, which show a strong adherence to FDC ("rosettes" in cytocentrifuge preparations). Generally, T cells and macrophages are absent in the preparations. Additional purification procedures involve enrichment/depletion on the basis of cell surface markers (FREEDMAN et al. 1992; PARMENTIER et al. 1991a; SCHMITZ et al. 1993; STAHMER et al. 1991). The FDC-containing suspensions can be cultured for a short period, after which the cells are generally lost from culture (CLARK et al. 1992; TSUNODA et al. 1990) or can no longer be detected by immunocytochemistry. Therefore, most studies on FDC in suspension or after culture require an analytic readout at the single cell level. In our hands, FDC have been kept in culture for 3–5 days, using supplementation of the culture with the supernatant of mitogen-activated blood mononuclear cells.

Despite technical complications, studies on FDC in suspension have yielded valuable information, e.g., on the immunologic phenotype and biologic function of the cells. Our data on the immunologic phenotype of FDC isolated from human tonsils are presented in Table 1. The most specific identification of the cells is by the monoclonal antibodies anti-DRC-1 and Ki-M4. These antibodies are useful when applied in tissue section analysis; however, in cell suspension the antibodies also label B lymphocytes (about half of the blood B lymphocytes and almost all tonsillar B lymphocytes; PARMENTIER et al. 1991a). This marker expression apparently does not reflect the presence of exogenously acquired Ki-M4 or DCR-1 antigen, as Epstein-Barr Virus (EBV)-induced B lymphocyte cell lines also show immunolabeling. For some other markers, GERDES and FLAD

Table 1. Immunologic Phenotype of human tonsillar follicular dendritic cells (FDC)

Immunophenotype	Cell suspension	Tissue section
CD1	–	–
CD2	±	±
CD3	–	–
CD4	±[a]	±
CD5	–	±
CD8	–	–
CD22	++	++
CD25	–	–
CD30	–	–
CD37	±	+
CD35 (CR1,C3bR)	++	++
CD21 (CR2,C3dR)	++	++
CD11a (LFA–1) 1	–	–
CD11b (CR3, C3biR)	++	++
CD11c	+	+
CD18	+	+
anti–DRC–1/Ki–M4	++	++
CD64 (FcγRI)	–	±
CD32(FcγRII)	+	±
CD16(FcγRIII)	±	+
CD54(ICAM–1)	++	++
VCAM–1	++	++
CD49a–f(VLA–1–6)	–	–
sIgG	++	++
sIgM	++	++
sIgD	±	±
Clq	++	++
C3b	+	++
Macrophage[b]	±	±
S–100[c]	–	–

Data from JOLING et al. (1993, 1994), KOOPMAN et al. (1991), PARMENTIER et al. (1991a), TUIJNMAN et al. (1993), and VROOM et al. (1993).

CD, cluster of differentiation; CR, complement receptor; FcγR, receptor for Fc fragment of IgG; ICAM, intercellular adhesion molecule; LFA, Lymphocyte function–related antigen; sIg, surface immunoglobulin; VLA, very late antigen.

[a] Demonstrated by immunogold electron microscopy on about 20% of FDC

[b] Anti–macrophage antibody EBM11 (Dakopatts, Glostrup, Denmark).

[c] Marker for interdigitating dendritic cells.

(1992), HEINEN et al. (1984, 1985), PARMENTIER et al. (1991a), REYNES et al. (1985), SCHRIEVER et al. (1989, 1991), and SELLHEYER et al. (1989) have reported discrepant data.

A relevant surface marker in HIV-1 infection is the CD4 molecule, as this is the classical virus receptor (DALGLEISH et al. 1984; KLATZMANN et al. 1984). Using an antibody to the OKT4a epitope in immunogold electron microscopy, PARMENTIER et al. (1991a) have found CD4 expression on about 20% of isolated FDC; an antibody to the OKT4d epitope did not result in any labeling. Tissue section analysis is less informative in this respect, as CD4 antibodies generally result in weak reticular labeling of the germinal center. PARMENTIER et al. (1991a), REYNES et al. (1985), and SCHRIEVER et al. (1989) have demonstrated the presence of complement receptors on FDC, but there is ongoing debate on the presence of receptors for the Fc fragment of immunoglobulin (Ig)E (HARDIE et al. 1993; MAEDA et al. 1992; MASUDA et al. 1989) and that of IgG (FcγR; BONNEFOY et al. 1993; HEINEN et al. 1985; SCHRIEVER et al. 1989). The presence and extent of expression may be different between different species (GERDES and FLAD 1992). In humans, TUIJNMAN et al. (1993) have observed different results using various antibodies to the three FcγR classes. Two CD32 antibodies (IV.3 and KB61, against FcγRII) labeled isolated FDC in suspension, but only KB61 labeled cells in tissue sections. Three out of five CD16 antibodies (against FcγRIII) stained FDC in tissue sections, and two antibodies in this cluster labeled isolated FDC. This variability may be ascribed to the low density of FcγR on FDC and shielding of the receptors by immune complexes. In addition to their role in immune complex binding, Fc receptors have been claimed to play a role in the processing of B lymphocytes in the germinal center (BONNEFOY et al. 1993).

The presence of complement receptors and receptors for Fcγ is logical in view of the trapping of antigen in the form of immune complexes in the germinal center. These receptors may function to bind the complexes to the surface of FDC. Animal studies have in fact indicated a crucial role of complement activation and presence of Fc receptors in the antibody response in germinal centers (MAEDA et al. 1988; VAN DEN BERG et al. 1992; YAMAKAWA and IMAI 1992; ZWIRNER et al. 1989).

In addition to the formation of antigen–antibody complexes, adhesion molecules are also involved in the interaction between FDC and B cells (CLARK et al. 1992; KOSCO et al. 1992; PETRASCH et al. 1992; REE et al. 1993; RUCO et al. 1992). This has been documented in vitro using antibodies in cell adhesion assays for the intercellular adhesion molecule (ICAM)-1/lymphocyte function-associated antigen (LFA)-1 and the vascular cell adhesion molecule (VCAM)-1/very late antigen (VLA)-4 combination (FREEDMAN et al. 1990, 1992; KOOPMAN et al. 1991; LINDHOUT et al. 1993; LOUIS et al. 1989). In a detailed study using antibodies to various domains on the ICAM-1 molecule, JOLING et al. (1994) demonstrated that domain 5 of the molecule is involved in FDC–B cell interaction and not in homotypic B–B cell interaction, while the first domain of the molecule is involved in both FDC–B cell and B–B cell interaction. This suggests the presence of at least three different interactive pathways between FDC and B lymphocytes.

FDC are a heterogeneous population when examined at the electron microscopic level. Based on the content of cell organelles, appearance of cellular extensions, presence of electron-dense (immune complex) deposits, and occurrence of intermediate filaments, RADEMAKERS (1992) has differentiated seven types of FDC in the human tonsil (Table 2, Fig. 2). These features suggest a gradual spectrum in differentiation, activation, and regression among these FDC types; this forms the basis of the designations used in Table 2. The spectrum ranges from primitive to a low stage of differentiation for FDC types 1–3, to differentiated types 4 and 5, and regressive changes for types 6 and 7. These types of FDC show a preferential location in different parts of the germinal center of tonsils. FDC types with a low extent of differentiation occur mainly in the dark zone, and secretory active types 4 and 5 predominate in the light zone. This preferential localization of different FDC types suggests that the dark and light zones each have a specific microenvironment of FDC. This underlines the different support given by FDC for proliferation and differentiation of B lymphocytes in the different germinal center types of compartments.

Table 2. Ultrastructural features of follicular dendritic cell (FDC) types in germinal centres

FDC type	Designation	Ultrastructural characteristics
Type 1	Primitive	Very scarce cell organelles; when present, poorly developed. Filamentous cytoplasmic matrix. No villous extensions or electron-dense (immune complex) deposits
Type 2	Undifferentiated	Stellate cells with few organelles; polyribosomes. Submembraneous intermediate filaments. No villous extensions covered with electron-dense (immune complex) deposits
Type 3	Intermediate	Rounded cells with moderately developed rough endoplasmic reticulum (RER) and inconspicuous Golgi area. Some plumb cytoplasmic extensions having no electron-dense (immune complex) deposits. Submembraneous filament condensations
Type 4	Differentiated	Rounded cells with well-developed RER and Golgi systems. Some submembraneous filament condensations. Tiny villous plasma membrane protrusions with electron-dense (immune complex) deposits
Type 5	Secretory	Like type 4, but with a larger amount of cytoplasm. RER, Golgi area, and villous plasma membrane web extremely well developed. Many Golgi-associated vesicles. Electron-dense (immune complex) deposits present
Type 6	Regressive, pale	Elongated cells with large proportions of cytoplasm. Both RER (dilated) and Golgi area are inconspicuous. Lysosomes are present. Moderate density of villous extensions with electron-dense (immune complex) deposits that are partly engulfed by broad FDC extensions
Type 7	Regressive, dark	Stellate cells with electron-dense cytoplasmic and nuclear matrix. Dilated RER; inconspicuous Golgi area; clusters of free ribosomes present. Villous extensions and dense deposits not regularly present

Data from RADEMAKERS (1992). For illustrations see Figs. 2, 4 and 5.

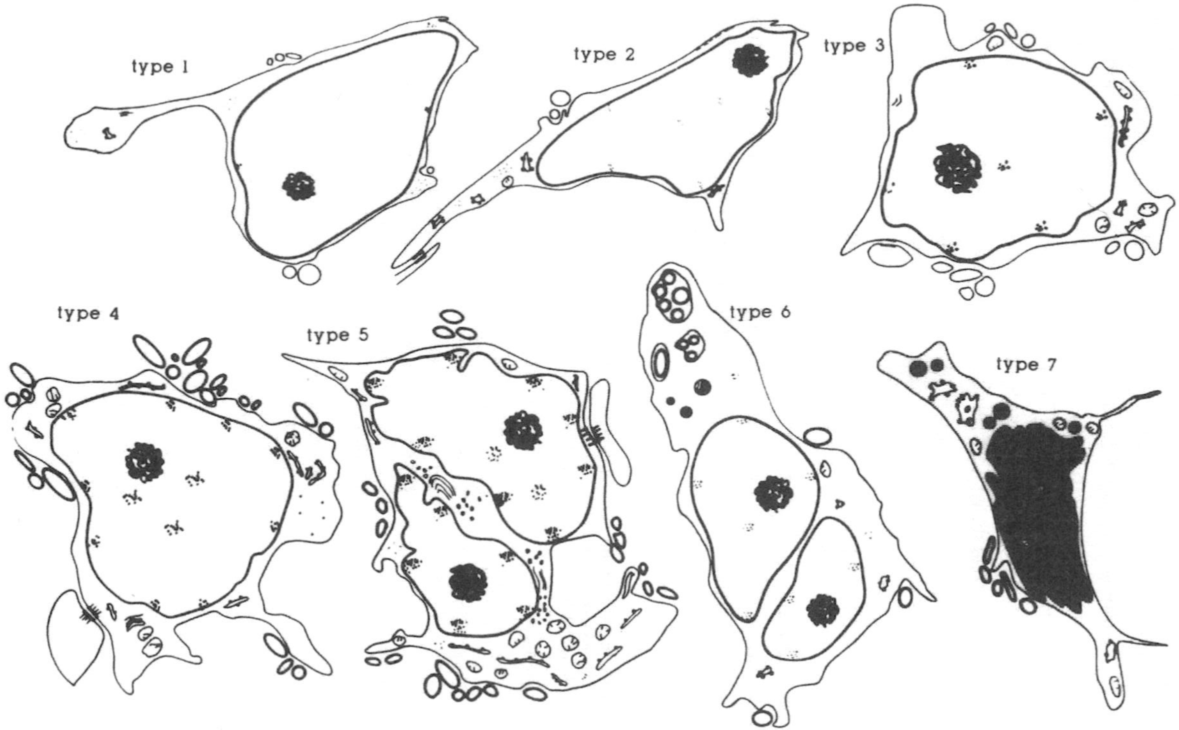

Fig. 2. Follicular dendritic cell (FDC) types in the germinal center of lymphoid tissue in humans. The description of these cell type is given in Table 2

3 Histologic Features of Lymph Nodes After Human Immunodeficiency Virus Type 1 Infection

Similar to the observations in conventional histology, major abnormalities in the follicles are observed in immunohistologic evaluation for markers of leukocytes and stromal cell types (Baroni and Uccini 1990; Biberfeld et al. 1985; Garcia et al. 1986; Janossy et al. 1985; Piris et al. 1987; Porwit et al. 1989; Rácz et al. 1986; Schuurman et al. 1985, 1988a; Wood 1990). In normal follicles the labeling pattern by anti-DRC-1 and Ki-M4 follows a confluent dendritic pattern of round- to oval-shaped germinal centers. However, in hyperplastic follicles after HIV-1 infection, the stroma visualized by anti-DRC-1 or Ki-M4 shows indentations and fragmentations (Fig. 3a). These are associated with the histologic appearance described above, but often are more easily seen in immunohistolology than in conventional histology. The framework formed by FDC shows the immunophenotype of normal germinal centers, e.g., the presence of immunoglobulins (IgG and IgM) and complement components and the virtual absence of IgD immunolabeling. B lymphocytes are present, as are also solitary $CD4^+$ T cells and macrophages. As in normal follicles $CD57^+$ cells (presumed natural killer cells) are present, but we have not found $CD8^+$ T cells (cytotoxic T cell phenotype) in the FDC-containing part of the follicle. However, Baroni and Uccini (1990), Baroni et al. (1988), Paravicini et al. (1989), Porwit et al. (1989), and Rácz et al. (1990), have reported the infiltration of germinal centers by $CD8^+$ cells.

The areas of indentations and fragmentations where there is no labeling by anti-DRC1 or Ki-M4 show lymphocytes with the phenotype of mantle cells, e.g., B lymphocytes expressing IgM and IgD, T lymphocytes of CD4 or CD8 phenotype (Schuurman et al. 1985, 1988a). This phenotype is in accordance with the cytology of the cells mentioned above, i.e., mainly comprising small-sized lymphocytes. Thus, the areas of indentation and fragmentation of the follicle should not be considered as part of the germinal center, but rather as representing the original follicle mantle (B cells) or interfollicular area. The stroma of the fragmented and indented areas has not been characterized thus far.

The interfollicular areas in hyperplastic lymph nodes show (immuno) histologic characteristics that are not very different from those in the normal lymph node. Schuurman et al. (1985) have noted in this respect that the CD4 to CD8 ratio within the lymph node T lymphocyte population is not as low as that within the blood T lymphocyte population. Recently, Rosenberg et al. (1993) made a similar observation in macaques after infection with simian immunodeficiency virus (SIV). A striking feature is the accumulation of $CD1^+$ cells with a dendritic morphology, presumably representing recently immigrated interdigitating dendritic cells presenting antigen to ($CD4^+$) T cells (Schuurman et al. 1985). High proportions of these cells, which strongly express HLA class II molecules, are found in other conditions as well, e.g., in some types of (congenital) immunodeficiency with defective T cell reactivity and in dermatopathic lymphadenopathy. In lymph nodes after HIV-1 infection, this phenomenon may

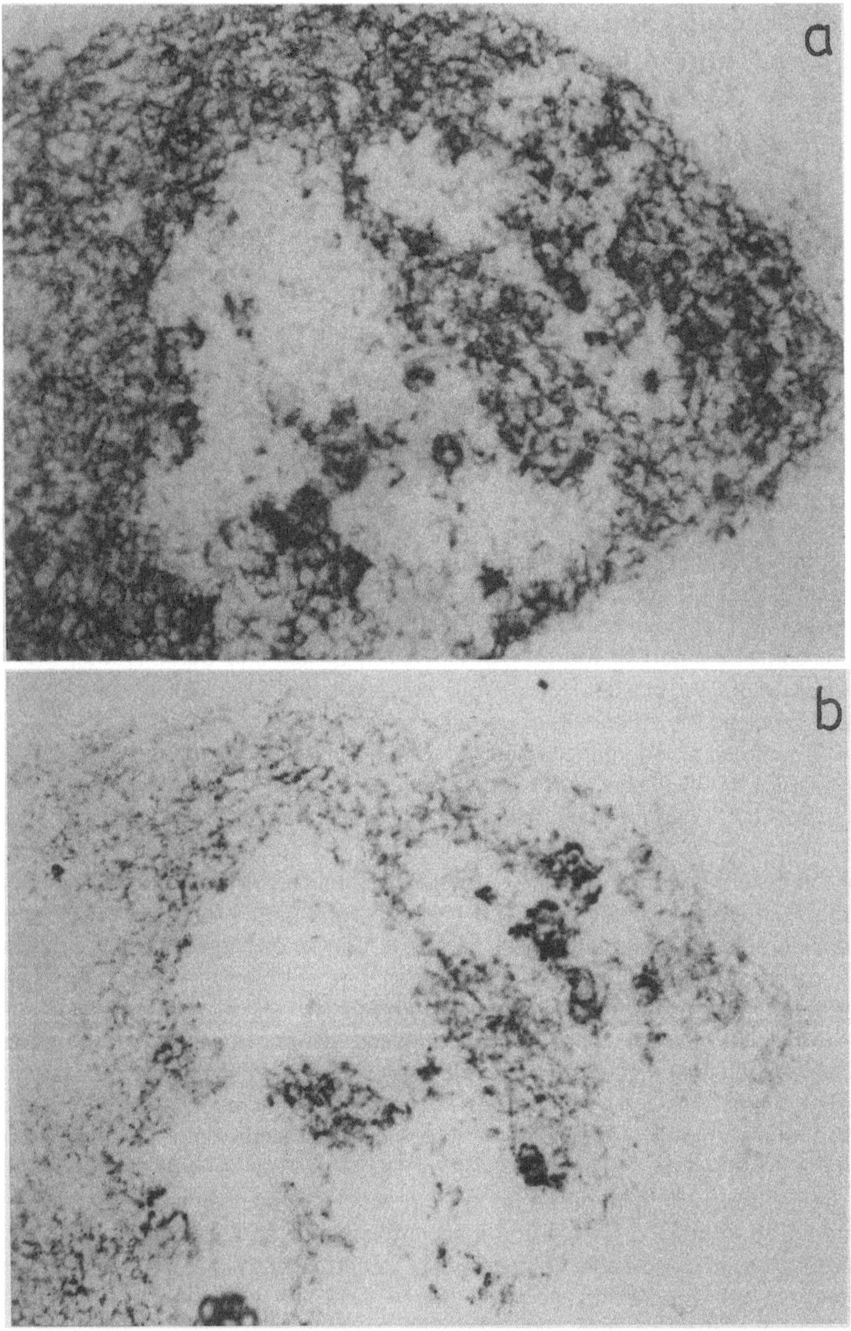
a
b

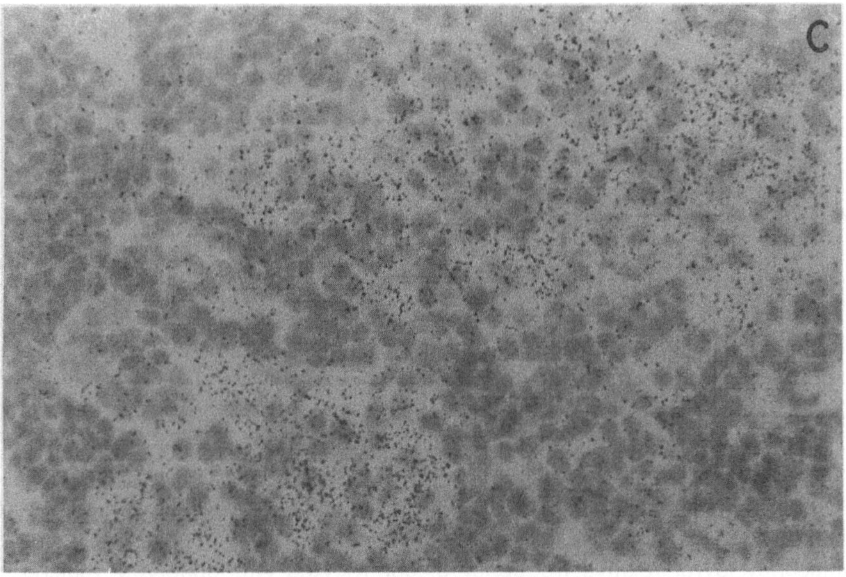

Fig. 3 a–c. Immunohistology of lymph node, follicular hyperplasia after human immunodeficiency virus (HIV)-1 infection. **a** Anti-DRC-1 labeling, showing indentation and fragmentation of the dendritic staining pattern. x 250. **b** Anti-HIV-1 gag immunolabeling of the same area in an adjacent section, showing a similar staining pattern. This indicates the presence of HIV-1 gag protein concentrated on the surface of the FDC. x 250. Frozen tissue section, immunoperoxidase immunolabeling, without counterstaining. **c** In situ hybridization for the presence of HIV-1 RNA, showing a diffuse presence of a hybridization signal over the area in the follicle where FDC are present. x 400. Frozen tissue section, hybridization with a ^{35}S-radioactive RNA probe complementary to HIV-1 RNA, autoradiographic detection, hematoxylin counterstain. The illustration in **c** is from a different lymph node than that in **a** and **b**

indicate the continuous exposure of antigen to T lymphocytes that have lost the capacity of adequate responsiveness.

The presence of $CD8^+$ cells in the germinal center, although in our experience mainly restricted to areas of indentations and fragmentations, has been related to the cytotoxic action of these cells, presumably associated with reactivity to virally infected cells and thus resulting in tissue destruction (Laman et al. 1989). Cytotoxic cells can be identified by an antibody to serine esterase granzyme B, which together with perforin is expressed in the granules of cytotoxic T cells and natural killer cells (Masson and Tschopp 1987; Peters et al. 1991). Parmentier et al. (1991b) found granzyme B immunolabeling in lymph nodes from HIV-1-infected patients, but not in higher proportions than in lymph nodes from non-HIV-1-infected patients. Granzyme B-positive cells were found only in low proportions in the interfollicular areas, and not in germinal centers. Joling et al. (1992) have made essentially similar observations in lymphoid tissues from cynomolgus monkeys after infection with SIV_{sm}, despite the relatively high density of $CD8^+$ cells. In contrast, Devergne et al. (1991) have demonstrated cells expressing mRNA of serine esterase granzyme B in lymphoid follicles, indicating the presence of cytotoxic cells. This discrepancy between immunohistochemical and hybridohistochemical data may be related to the specificity of the methods. As far as protein detection is concerned, it can be speculated that cytotoxic cells, if present, have lost their cytotoxic granules or that the protein is diffused through the tissue, reaching a local concentration below the detection level of immunohistochemistry. Alternatively, the discrepancy may be related to a difference in study material between the two reports. As the simultaneous detection of mRNA and protein in the same sample or tissue section has not been reported thus far, the issue concerning cytotoxic cells in the germinal center mediating tissue destruction cannot be unequivocally solved.

In the electron microscopic evaluation of FDC in tissue sections, Rademakers et al. (1992a) have used the FDC typing mentioned above. This was done for whole follicles, because lymph node follicles did not manifest a subdivision into light and dark zones. Cluster analysis of the relative frequencies of lymphoid cell types, histiocytic cells, and FDC resulted in two main groups (Table 3); one cluster included all cases of control hyperplasia and six of fifteen HIV-1-infected cases, and the other one included the other nine HIV-1-infected cases. In this second cluster a significantly higher relative frequency of lymphoid blast subtypes and a significantly lower relative frequency of centrocytes was observed, making a blast to centrocyte ratio of 1.6 in the second cluster versus 0.5 in the first one (Fig. 4). The subsequent evaluation of these clusters for the distribution of FDC types showed significant differences; the second cluster manifested a significantly lower frequency of differentiated types 4 and 5 and a significantly higher frequency of regressive types 6 and 7 (Fig. 5). Thus, FDC in the major part of lymph nodes from HIV-1 infected patients do not manifest a high extent of differentiation, and they show signs of regression. In the light and dark zone of

Table 3. Mean relative frequency (%) of lymphoid cell types and follicular dendritic cell (FDC) types in germinal centers of lymph nodes from (HIV)-1 infected patients (n = 15) and controls (n = 8), grouped according to clustering

Cell type	Cluster 1	Cluster 2
Number of patients	14	9
Number of HIV-1 infected patients	6	9
Lymphoid blastoid cells	26.3	48.8
Centrocytes	49.2	30.9
FDC types		
Type 1	0.1	0.4
Type 2	5.7	7.1
Type 3	23.6	22.9
Type 4	33.0	20.1
Type 5	17.8	11.5
Type 6	11.6	21.6
Type 7	8.9	15.8

[1]Data from Rademakers et al. (1992a). Clustering was done on the basis of the relative frequencies of lymphoid cell types (blastoid cells, cleaved blasts, immunoblasts and centroblasts, centrocytes, centroplasmacytoid cells, lymphocytic multilobated cells, and plasma cells) and that of histiocytic cells and FDC. The main discriminative parameters in this analysis were the blastoid cells and the centrocytes (data presented in the table). Subsequently, FDC types in the clusters were assessed.

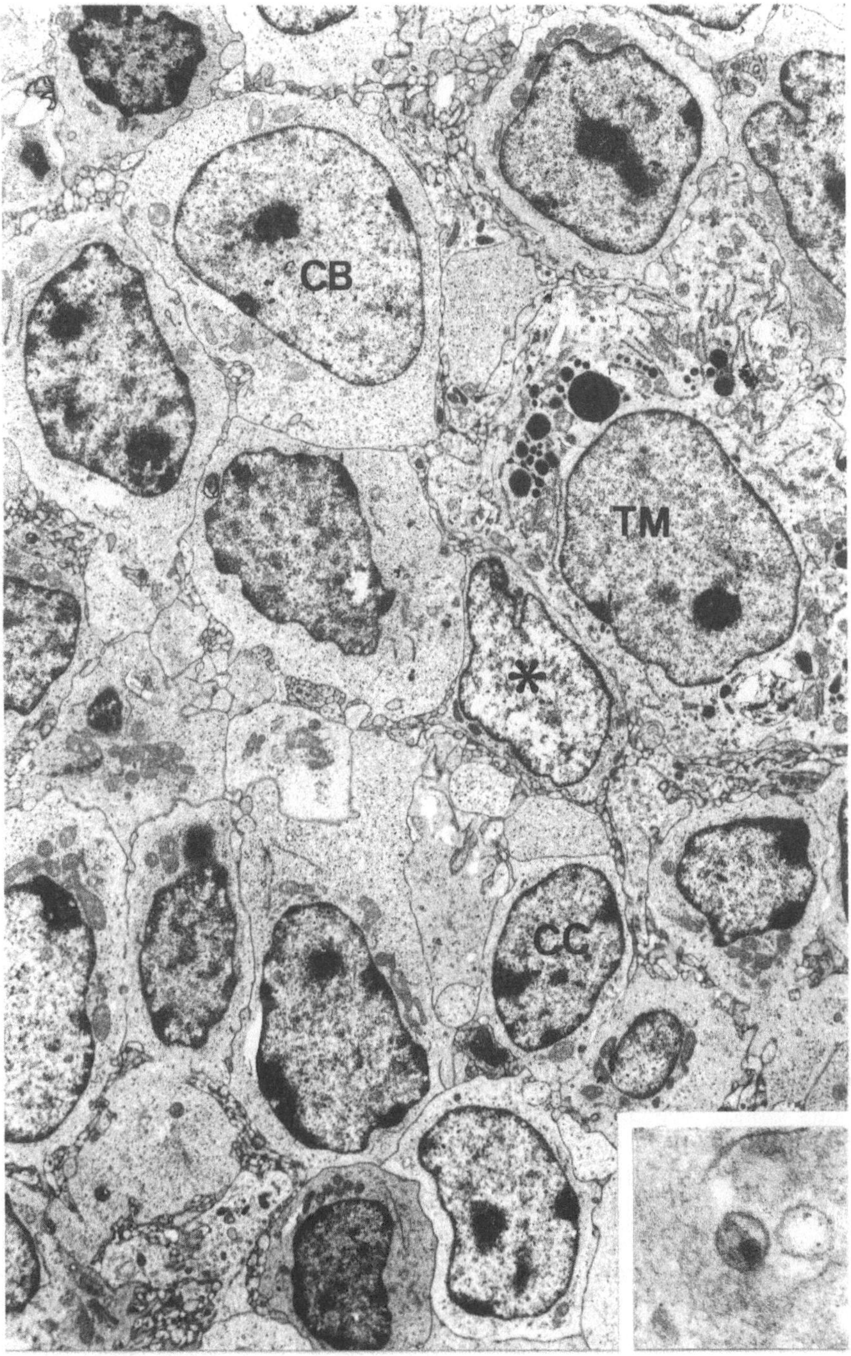

Fig. 4. Electron micrograph showing an overview of a germinal center in a lymph node from a human immunodeficiency virus (HIV)-1-infected patient (cluster 2, Table 3). The main lymphoid population comprises centroblasts (*CB*), and few centrocytes (*CC*) are present. There is a follicular dendritic cell (FDC, type 4; *asterisk*) in close apposition with a tingible body macrophage a (*TM*). x 4000. *Inset*, HIV-1 virus particle. x 80 000

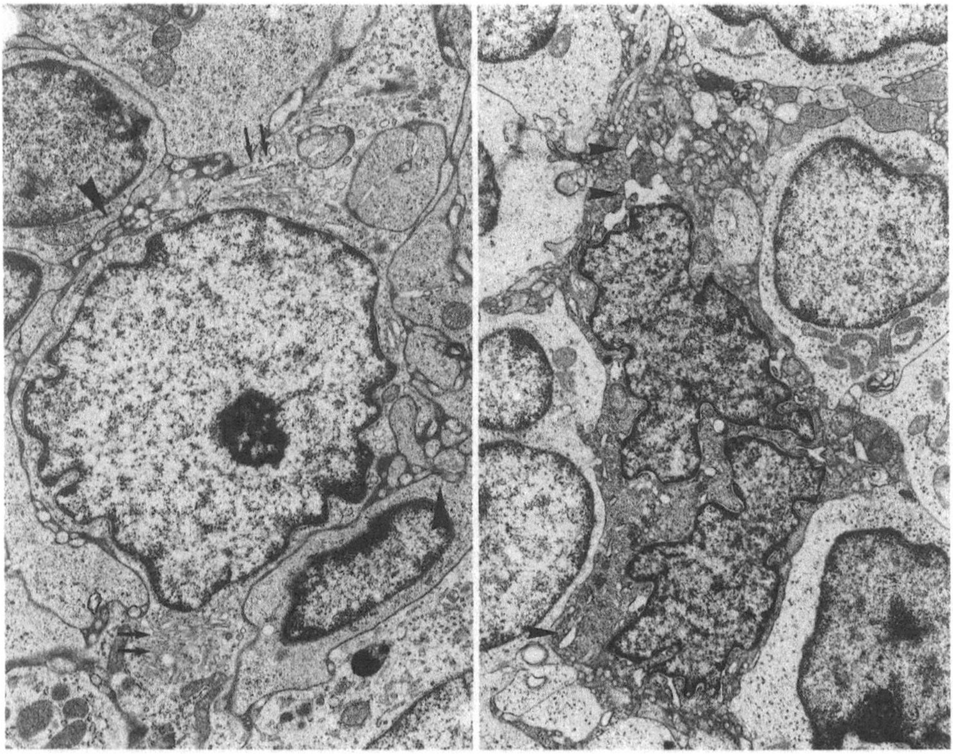

Fig. 5. *Left* Electron micrograph of a type 5 follicular dendritic cell (FDC) with large cytoplasm containing an extensive Golgi area with several Golgi complexes (*arrows*). There are many villous extensions with electron-dense deposits (*arrowheads*). x 5850. *Right* Binucleated type 7 FDC with electron-dense cytoplasmic and nuclear matrix. Cisterns of rough endoplasmic reticulum are dilated (*arrowheads*). There are villous extensions with electron-dense deposits visible. x 5850

tonsillar germinal centers this peculiar cell combination of regressive FDC types and a high blast to centrocyte ratio does not occur. A high blast to centrocyte ratio is present only in the dark zone, in combination with the primitive/undifferentiated FDC types 1-3. The observation in the second cluster of lymph nodes is indicative of an inadequate support by FDC for germinal center B lymphocyte differentiation after HIV-1 infection.

JOLING et al. (1992) have described similar changes in spleen and lymph nodes from cynomolgus monkeys after experimental infection by SIV_{sm}. A remarkable observation was the presence of paracrystalline arrays resembling so-called tubuloreticular structures in FDC with the ultrastructural features of undifferentiated fibroblastoid cells. KOSTIANOVSKY et al. (1983), ONERHEIM et al. (1984), and SCHIAFFINO et al. (1986) have documented such structures in cells after retroviral infection, including lymphocytes and endothelial cells, but not in FDC. GRIMLEY et al. (1985) and LUU et al. (1989) have associated the presence of tubuloreticular structures with the synthesis of α-interferon by the cell itself or with the reaction of the cell to the presence of α-interferon in the local environment. Their presence in FDC indicates that these accessory cells in the germinal center contribute to the response to viral infection.

Based on these histologic observations, the hypothesis has been put forward that FDC in lymphoid tissue go into a process of differentiation and regression after HIV-1 infection. They are then destroyed and disappear, and indentations and fragmentations of the FDC meshwork emerge in the germinal center. Due to the absence of the stromal component, lymphocytes from the follicle mantle and interfollicular area are able to invade these areas, resulting in the histologic appearance described above. This change is not pathognomonic for HIV-1 infection, as it is found in other conditions as well. In our experience, indentations and fragmentations are frequently encountered in germinal centers in mucosa-associated lymphoid tissue, e.g., tonsils and plaques of Peyer. Lymph nodes can also show fragmentations and indentations of germinal centers in conditions of systemic autoimmune diseases, such as systemic lupus erythematosus, and in some other viral infections. Apparently, a number of initiating events can underly such germinal center abnormalities. These abnormalities are very pronounced after HIV-1 infection and can occur in lymphoid tissue from SIV_{sm}-infected cynomolgus monkeys. It is therefore likely that these immunodeficiency viruses have a causative role in the evolution of histologic abnormalities, i.e., the destruction of the FDC meshwork. This suggestion can be explained in two ways: on the one hand by the cytopathic action by the virus itself, and on the other hand by the immune response of the host to the virus. As stated above, there is no firm indication that host cytotoxic cells to the virus contribute to follicle germinal center destruction. In the following section we focus on the possible FDC destruction by the virus itself.

4 In Situ Identification of Human Immunodeficiency Virus Type 1

Viral particles are easily identified in the germinal center of lymphoid tissue from patients after HIV-1 infection (Armstrong and Horne 1984; Cameron et al. 1987; Le Tourneau et al. 1986; O'Hara et al. 1988; Rácz et al. 1990; Rademakers et al. 1992a; Tenner-Rácz 1988; Tenner-Rácz et al. 1985, 1989; Warner et al. 1984) or monkeys after SIV infection (Joling et al. 1992; O'Hara et al. 1988; Ringler et al. 1989; Tenner-Rácz et al. 1985). Virus particles, 80–120 nm in diameter with a large electron-dense core, occur between villous extensions of FDC (Fig. 4, inset). The viral particles presumably are present in the form of antigen–antibody complexes on the surface of FDC. In our experience, budding of viral particles was not detectable, but Rácz et al. (1990) and Tenner-Rácz et al. (1989) have observed signs of viral replication in FDC. Rademakers et al. (1992a) were unable to identify budding phenomena or ultrastructural indications of syncytium formation that are observed after in vitro interaction of virus with $CD4^+$ cells.

Baroni and Uccini (1990), Baroni et al. (1988), Biberfeld et al. (1988), Garcia et al. (1986), Parmentier et al. (1990), Pekovic et al. (1987) Rácz et al. (1990), Schuurman et al. (1988a), Tenner-Rácz et al. (1986), and Wood (1990) have identified viral proteins using specific antibodies, either to the structural proteins gag and env or to the regulatory proteins tat, rev, and nef. In lymph nodes the highest labeling for structural proteins gag and env has been demonstrated in the germinal center (Fig. 3b). The staining colocalized with the labeling for FDC by anti-DRC-1 or Ki-M4 and with immunoglobulin and complement components. This was best demonstrated in follicles with indentations and fragmentations: the location where there was no labeling for FDC antigen was also negative for HIV-1 antigen (compare Fig. 3a and b). In addition, FDC isolated from a lymph node of a HIV-1-infected patient (Parmentier et al. 1990) and from lymphoid tissue of SIV_{sm}-infected monkeys (Joling et al. 1992) showed an intense immunolabeling for HIV-1 or SIV viral antigens. This immunolabeling is concordant with the trapping of viral proteins in the FDC labyrinth, presumably in the form of immune complexes. This trapping points to the presentation of HIV-1 antigen to B cells, analogous to a normal humoral immune response to an exogenous antigen. It is not clear whether this trapping of viral protein has any consequence for germinal center destruction.

The highest immunolabeling in the FDC-containing germinal center was observed for gag protein, exceeding that for env gp41 protein. For unknown reasons, we were unable to visualize env using an anti-env gp120 antibody (Parmentier et al. 1990; Schuurman et al. 1988a, 1989). Outside germinal centers, immunolabeling was found for solitary cells, in some cases also on the endothelium. Some of these solitary cells in the interfollicular area showed the morphology of interdigitating dendritic cells and were strongly positive for HLA class II antigen in two-color immunolabeling. In contrast to the germinal center, the intensity of labeling of these cells for env gp41 generally was higher than that for gag proteins. It is not clear whether this differential intensity of immunolabeling product for gag and env proteins between FDC in the germinal center and interdigitating dendritic cells in the interfollicular area is relevant for the mode of HIV-1 presentation to different lymphocyte subsets at these different sites.

The specificity of immunolabeling of lymph nodes has been investigated, including the analysis of lymph nodes with follicular hyperplasia from uninfected patients. It became apparent that some anti-gag antibodies result in immunolabeling of tissues from uninfected persons, e.g., epithelial cells in thymus and tubules in the kidney (Parravicini et al. 1988; Schuurman et al. 1988b, 1989). This apparent cross reactivity is relevant for diagnostic immunohistology: one should be very careful in interpreting HIV-1 immunolabeling on tissue sections. We therefore recommend that immunohistochemistry should not be used to diagnose HIV-1 infection, but rather that its use be restricted to studies on pathogenesis or prognosis, for example.

The basis of the cross-reactivity observed is still a matter of speculation. Apparently, epitopes recognized by (monoclonal) anti-HIV-1 antibodies do occur in normal human tissue. Beretta et al. (1987), Purcell et al. (1989), Query and

KEENE (1987), and SCHUURMAN et al. (1988b) have described the presence of cross-reactive epitopes or homologous regions. Examples are human HLA class II molecules and HIV-1 env protein (GOLDING et al. 1988; PUGLIESE et al. 1992) and the Arg-Gly-Asp (RGD) sequence on human cell adhesion molecules and HIV-1 tat (regulatory) protein (BRAKE et al. 1990). For HIV-1 gag (NAYLOR et al. 1987; SCHUURMAN et al. 1988b) and env (NGUYEN and SCHEVING 1987) there is homology with the thymic hormone thymosin-α_1, which has been claimed to underly the virus-neutralizing capacity of antithymosin antisera (SARIN et al. 1986). RITTER et al. (1987) have disputed this claim. This aside, thymic epithelium, being the source of thymosin, apparently expresses molecules sharing epitopes with retroviral antigens. For instance, COHEN-KAMINSKY et al. (1987) and HAYNES et al. (1983) have successfully applied an antibody to p19 gag antigen of human T lymphotropic virus I in the detection of thymic epithelium. A second explanation comes from the existence in the genome of endogenous proretroviral or retrovirus-like sequences that have been documented in humans by BRACK-WERNER et al. (1989), LARSSON et al. (1989), and WILKINSON et al. (1990) and in other species such as the baboon by COHEN et al. (1981) and the mouse by KOZAK (1985).

For various antibodies to HIV-1 regulatory proteins tat, rev, and nef, PARMENTIER et al. (1992) have observed an even higher degree of cross-reactivity than for antibodies to HIV-1 structural proteins. In lymphoid tissue from uninfected patients, anti-tat antibodies from two different sources labeled endothelium, and an antibody to rev labeled histiocytes. FDC were labeled by one out of three antibodies to HIV-1 nef. Lymph nodes from HIV-1-infected individuals manifested a similar labeling, in some cases more intensely than those from uninfected subjects. Also, in conditions of local inflammation, e.g., skin in patients with atopic dermatitis, there may be upregulation of the cross-reactive epitopes. SCHUURMAN et al. (1993) have found this for keratinocyte labeling by anti-rev antibody and for Langerhans cells by anti-tat antibody. Thus, it is not possible to study the regulation of HIV-1 expression in situ using such antibodies to HIV-1 regulatory proteins.

HIV-1 RNA can be detected on tissue sections by in situ hybridization. Using optimal (autoradiographic) detection with ^{35}S-labeled probes (complementary to HIV-1 gag/pol RNA or to env RNA) and autoradiographic detection, a low density of cells has been observed at scattered locations in lymph nodes from HIV-1-infected individuals (BARONI et al. 1988; PARMENTIER et al. 1990; SCHUURMAN et al. 1988a, 1989). We were unable to document differences between follicles and interfollicular areas in the density of cells expressing HIV-1 RNA, but BIBERFELD et al. (1986, 1988), BURKE et al. (1993), and RÁCZ et al. (1990) have described a higher density of solitary RNA-positive cells in follicles than in interfollicular areas. PREVOT et al. (1989) and SPIEGEL et al. (1992) have observed a confluent low-intensity hybridization signal in the FDC-containing area in germinal centers, but not at other locations in the tissue section. We have also observed this in some cases (Fig. 3c). This phenomenon has been interpreted as reflecting a state of HIV-1 infection of the stroma, i.e., of FDC. This conclusion strictu sensu is not allowed, as the signal may reflect a high concentration of virus at this location, i.e., the presence of large numbers of viral particles at extracellular sites entrapped in the FDC labyrinth.

PARMENTIER et al. (1990) have also detected HIV-1 RNA in FDC isolated from a lymph node of a HIV-1-infected patient. The cytocentrifuge preparation subjected to HIV RNA hybridization manifested a signal over the cytoplasm of FDC-like cells. This has also been interpreted as reflecting HIV-1 infection and active virus production by the cells. This interpretation is subject to the same comment as that on the low-intensity hybridization signal over FDC in tissue sections, as it can be the reflection of large numbers of viral particles bound to the isolated cells. Due to the low yield of cells from the lymph node, we were unable to address this question in detail, e.g., by electron microscopy.

Another approach to address this question is to perform infection experiments in vitro. Such studies are hampered by the impossibility of obtaining pure FDC suspensions without contamination by other cells and thus require analysis at the single cell level. JOLING et al. (1993) have studied the interaction between HIV-1 and FDC using HIV-1 particles conjugated to fluorescein isothiocyanate (FITC) by flow cytometry (Table 4). The highest extent of binding was observed when the incubation medium was supplemented with fresh (unheated) serum from control donors, and an additional effect was observed for heated serum from HIV-1 patients containing anti-HIV-1 antibodies. This data was confirmed by fluorescence microscopy of the cells after processing to cytocentrifuge preparations. The factor in fresh unheated serum mediating the binding of HIV-1/FITC to FDC proved to be complement component C3, as shown

Table 4. Fluorescence signal of follicular dendritic cells (FDC) or blood mononuclear cells after incubation with fluorescein isothiocyanate (FITC)–labeled human immunodeficiency virus (HIV)-1 particles

Serum	Mean fluorescence	Quenching[a]	
		–	+
Follicular dendritic cells			
Background, no HIV-1/FITC	5.3	4.1	3.9
Control heated serum	8.2	5.9	6.4
Control fresh serum	66.7	20.3	8.2
Heated fresh serum	n.d.	5.9	5.0
HIV-1 serum and control fresh serum	n.d.	34.9	9.7
C3-deficient serum	8.3	n.d.	n.d.
C5-depleted serum	46.3	n.d.	n.d.
Blood mononuclear cells			
Control heated serum	n.d.	6.4	7.4
Control fresh serum	n.d.	57.5	29.0

Data from JOLING et al. (1993). FDC were incubated with FITC-conjugated HIV-1 particles for 30 min and thereafter fixed with paraformaldehyde. The binding was performed in the presence (20%) of various serum supplements that were either fresh or heated (56°C, 30 min) to destroy complement activity. The analysis was done by flow cytometry.

n.d., not done.

[a] Quenching to eliminate extracellular fluorescence was assessed by addition of trypan blue (–, not added; +, incubation with trypan blue).

by experiments using serum from a patient with deficiency of the third complement component and serum depleted of complement component C5. Thus, complement component C3 mediates binding of HIV-1/FITC to FDC. This is supported by the presence of complement receptors on FDC. BAKKER et al.(1992), BOYER et al.(1991), JUNE et al. (1991), and MONTEFIORI et al. (1993) have demonstrated that complement reinforces the binding of HIV-1 to various cell types. In addition, HIV-1 is able to activate the complement cascade, especially the classical pathway (DIERICH et al. 1993; EBENBICHLER et al. 1991; REISINGER et al. 1990; SÖLDER et al. 1989; SPEAR et al. 1990). Anti-HIV-1 antibodies have a role in virus binding and subsequent cellular infection in a number of cell types (BOYER et al. 1991; HOMSY et al. 1989; JOUAULT et al. 1989; JUNE et al. 1991; MONTEFIORI et al. 1990; ROBINSON et al. 1988, 1989; TAKEDA et al. 1988). This is not the case for HIV-1/FITC binding to FDC, which is presumably related to the low expression of FcγR by the cells. Experiments using fluorescence quenching showed a strong reduction of the fluorescence signal in the presence of trypan blue, indicating that most fluorescence was present extracellularly (Table 4). For blood mononuclear cells after HIV-1/FITC binding, quenching resulted in a smaller reduction of the fluorescence signal. Thus, under the conditions of the experiments, HIV-1/FITC remains bound on the surface of FDC whereas it is internalized by blood mononuclear cells (mainly monocytes). This result was confirmed by immunogold electron microscopy with anti-FITC and anti-HIV-1 gag antibody, in which most label was found on the FDC surface (JOLING et al.1993). These data are in accordance with those of STAHMER et al.(1991), who have demonstrated in vitro infection of FDC by HIV-1 after binding in a CD4-independent manner. Apparently, the binding of HIV-1 to FDC does not involve the classical interaction between HIV-1 env protein and the CD4 receptor molecule on the cell surface, as described for other cell types (LIFSON et al. 1986; SODROSKI et al. 1986; YOFFE et al. 1987). CHEHIMI et al. (1993) have reported a similar CD4-independent pathway in infection of blood dendritic cells.

5 Conclusions

CD4⁺T-helper lymphocytes and subsequently macrophages were originally proposed as the most important target cells for HIV-1 infection and virus spread through the body. An important role has also been proposed for dendritic cells and Langerhans cells which have a main antigen-presenting function for T-helper cells in primary immune responses (CAMERON et al. 1992; LANGHOFF and HASELTINE 1992; LANGHOFF et al. 1991, 1993; MACATONIA et al. 1989, 1990; PATTERSON et al. 1991; RODRIGUEZ et al. 1991; TSCHACHLER et al. 1987). These claims were made neglecting the abberrations in the follicles of swollen lymph nodes in HIV-1 infected patients. It is now clear that the germinal center, in particular the stromal component or FDC, represents a third target in the infectious process (GERDES and FLAD 1992;

Fox and Cottler-Fox 1992). This is evidenced by the substantial presence of virus at this location as viral particles, viral RNA, or viral proteins. Immunohistochemical evaluation shows that the germinal center of lymphoid organs is almost the only place in the body where an abundant expression of viral protein occurs. It is thus tempting to conclude that the germinal center is the main reservoir of extracellular virus in the body. Lymphocytes and cells of the macrophage series may represent the main reservoir of intracellular virus in this regard.

The relevance of the presence of virus in the germinal center can be illustrated by the results of a preliminary study by Joling et al. (1992) on SIV_{sm} infection in cynomolgus monkeys. Two animals received experimental treatment by dideoxyinosine in combination with azidothymidine or '3-fluorothymidine during the first 9 weeks after infection. FDC obtained at autopsy 25 weeks later showed a low intensity of SIVp28 immunoreactivity on cytocentrifuge preparations and no detectable immunoreactivity on tissue sections. An intense immunolabeling signal was observed on cells from three other animals who did not receive treatment. In addition, the treated animals showed the highest proportions of $CD4^+$ cells and the highest CD4 to CD8 ratio in peripheral blood. This data provides a first indication for an association between the effect of experimental antivirus treatment and the presence of virus on FDC in the germinal center. This correlates with data from Rosenberg et al. (1993) on the association between the extent of viral expression on FDC and the CD4 to CD8 T cell ratio in blood and lymph nodes in macaques after SIV infection. Embretson et al. (1993) and Panteleo et al. (1993a,b) have presented data on active processes of viral replication within lymphoid tissue and not in blood during the first phase after HIV infection.

FDC can be infected by HIV-1 as shown by in vitro infection experiments. It remains to be established whether this actual infection of FDC is important or whether the sole binding and persistence of HIV-1 particles in the extracellular space of the germinal center labyrinth represents the main reservoir for the infectious process. In this respect, our in vitro data indicate that FDC are able to bind virus and that viral particles subsequently internalize less efficiently in FDC than in blood mononuclear cells.

The changes in the lymphoid follicle are associated with a process of destruction. Indentation and fragmentation of the FDC-containing stromal meshwork already occurs in the follicular hyperplasia stage, and at the subcellular level regressive changes are observed in the major part of HIV-1-infected patients. The mechanism of this destruction is unknown. It is likely that the cause of this process is the virus itself and not the host response to the virus, because the virus is present in germinal centers and there is no clear indication of cytotoxic lymphocytes at this location. Following the reasoning mentioned above, a cytopathic action of extracellular virus seems more likely than actual cellular infection and intracellular processes leading to cell death. It is known that HIV-1 at the surface of cells can mediate cellular destruction in vitro via the process of syncytium formation. Such syncytium cells are rarely seen in situ and not in the germinal center. However, this neither proves nor disproves in vivo cell death by extracellular action, first because the process may occur very fast and thus is not visible in histologic slides, and second because the ex vivo histologic readout may be different from in vitro syncytium formation. Obviously, there is a need for adequate in vitro models to address these questions in detail.

If the processes inside germinal centers are accepted as crucial events in HIV-1 infection and disease development subsequent to systemic lymphadenopathy, we should be aware of the fact that these processes may involve other types of cellular interactions. In the classical route of infection, the interaction between the CD4 receptor molecule and the env protein of the HIV-1 plays an important role. However, the interaction between HIV-1 and FDC can occur in a CD4-independent pathway and requires the presence of complement component C3. The interaction between HIV-1 and complement is significant for the infectious process for a number of cells. DIERICH et al. (1993) have recently reviewed the ability of the virus to avoid complement-mediated lysis and to use complement components for its own benefit. In the cellular interaction inside germinal centers too, the ICAM-1/LFA-1 and VLA-4/VCAM-1 interactions may be important. Therefore approaches interfering with infectious processes inside germinal centers should focus on these molecular mechanisms rather than on the CD4-env gp 120 pathway.

Acknowledgment. The work reported in this manuscript was supported in part by the Durch Ministry of Health as part of the National Program on AIDS Research, Grant No. RGO/WVC 88-79/89005. The studies on cynomolgus monkeys after experimental SIV_{sm} infection were performed in collaboration with Dr. P. Biberfeld, Immunopathology Laboratory, Karolinska Institute, Stockholm, Sweden. The studies on in vitro interaction of HIV-1 with FDC were done in collaboration with Dr. L.J. Bakker, Dr. J.A.G. van Strijp, and Dr. J. Verhoef, Department of Clinical Microbiology, University Hospital, Utrecht, The Netherlands.

References

Armstrong JA, Horne R (1984) Follicular dendritic cells and virus-like particles in AIDS-related lymphadenopathy. Lancet 2: 370–372

Bakker LJ, Nottet HSLM, De Vos NM, De Graaf L, Van Strijp JAG, Visser MR, Verhoef J (1992) Antibodies and complement enhance binding and uptake of HIV-1 by human monocytes. AIDS 6: 35–41

Baroni CD, Uccini S (1990) Lymph nodes in HIV-positive drug abusers with persistent generalized lymphadenopathy: histology, immunohistochemistry, and pathogenetic correlations. Prog AIDS Pathol 2: 33–50

Baroni CD, Pezzella F, Pezzella M, Macchi V, Vitolo D, Uccini S, Ruco LP (1988) Expression of HIV in lymph node cells of LAS patients. Immunohistology, in situ hybridization, and identification of target cells. Am J Pathol 133: 498–506

Beretta A, Grassi F, Pelagi M, Clivio A, Parravicini C, Giovinazzo G, Andronico F, Lopalco L, Verani P, Butto S, Titti F, Rossi GB, Viale G, Ginelli E, Siccardi AG (1987) HIV env glycoprotein shares a cross-reacting epitope with a surface protein present on activated human monocytes and involved in antigen presentation. Eur J Immunol 17: 1793–1798

Biberfeld P, Porwit-Ksiasek A, Böttiger B, Morfeld-Månson L, Biberfeld G (1985) Immunohistopathology of lymph nodes in HTLV-III infected homosexuals with persistent adenopathy or AIDS. Cancer Res 45 [Suppl]: 4555S–4560S

Biberfeld P, Chayt KJ, Marselle LM, Biberfeld G, Gallo RC, Harper ME (1986) HTLV-III expression in infected lymph nodes and relevance to pathogenesis of lymphadenopathy. Am J Pathol 125: 436–442
Biberfeld P, Porwit A, Biberfeld G, Harper M, Bodmer A, Gallo R (1988) Lymphadenopathy in HIV (HTLV-III/LAV) infected subjects: the role of virus and follicular dendritic cells. Cancer Detect Prev 12: 217–224
Bonnefoy JY, Henchoz S, Hardie D, Holder MJ, Gordon J (1993) A subset of anti-CD21 antibodies promote the rescue of germinal center B cells from apoptosis. Eur J Immunol 23: 969–972
Boyer V, Desgranges C, Trabaud M-A, Fischer E, Kazatchkine MD (1991) Complement mediates human immunodeficiency virus type 1 infection of a human T cell line in a CD4- and antibody-independent fashion. J Exp Med 173: 1151–1158
Brack-Werner R, Leib-Mosch C, Werner T, Erfle V, Hehlmann R (1989) Human endogenous retrovirus-like sequences. Haematol Bloodtransf 32: 464–477
Brake DA, Debouck C, Biesecker G (1990) Identification of an Arg-Gly-Asp (RGD) cell adhesion site in human immunodeficiency virus type 1 transactivation protein, tat. J Cell Biol 111: 1275–1281
Burke AP, Benson W, Ribas JL, Anderson D, Chu WS, Smialek J, Virmani R (1993) Postmortem localization of HIV-1 RNA by in situ hybridization in lymphoid tissues of intravenous drug addicts who died unexpectedly. Am J Pathol 142: 1701–1713
Burns BF, Wood GS, Dorfman RF (1985) The varied histopathology of lymphadenopathy in the homosexual male. Am J Surg Pathol 9: 287–297
Cameron PU, Dawkins RL, Armstrong JA, Bonifacio E (1987) Western blot profiles, lymph node ultrastructure and viral expression in HIV-infected patients: a correlative study. Clin Exp Immunol 68: 465–478
Cameron PU, Freudenthal PS, Barker JM, Gezelter S, Inaba K, Steinman RM (1992) Dendritic cells exposed to human immunodeficiency virus type-1 transmit a vigorous cytopathic infection to CD4+ T cells. Science 257: 383–387
Centers for Disease Control (1986) Classification system for human T-lymphotropic virus type III/ lymphadenopathy associated virus infections. Morb Mort Wkly Rep 35: 334–339
Chehimi J, Prakash K, Shanmugam V, Collman R, Jackson SJ, Bandyopadhyay E, Starr SE (1993) CD4-independent infection of human peripheral blood dendritic cells with isolates of human immunodeficiency virus type 1. J Gen Virol 74: 1277–1285
Clark EA, Grabstein KH, Shu GL (1992) Cultured human follicular dendritic cells. Growth characteristics and interactions with B lymphocytes. J Immunol 148: 3327–3335
Cohen M, Rein A, Stephens R, O'Connell C, Gildman RV, Shure M, Nicolson MO, McAllister RM, Davidson N (1981) Baboon endogenous virus genome: molecular cloning and structural characterization of nondefective viral genomes from DNA of a baboon cell strain. Proc Natl Acad Sci USA 78: 5207–5211
Cohen-Kaminsky S, Berrih-Aknin S, Savino W, Dardenne M (1987) Immunodetection of the thymic epithelial p19 antigen in cultures of normal and pathologic human thymic epithelium. Thymus 9: 225–238
Dalgleish AG, Beverley PCL, Clapham PR, Crawford DH, Greaves MF, Weiss RA (1984) The CD4 (T4) antigen is an essential component of the receptor for the AIDS retrovirus. Nature 312: 763–767
Devergne O, Peuchmaur M, Crevon M-C, Trapani JA, Maillot M-C, Galanaud P, Emilie D (1991) Activation of cytotoxic cells in hyperplastic lymph nodes from HIV-infected patients. AIDS 5: 1071–1079
Dierich MP, Ebenbichler CF, Marschang P, Füst G, Thielens NM, Arlaud GJ (1993) HIV and human complement: mechanisms of interaction and biological implication. Immunol Today 14: 435–440
Ebenbichler CF, Thielens NM, Vornhagen R, Marschang P, Arlaud GJ, Dierich MP (1991) Human immunodeficiency virus type 1 activates the classical pathway of complement by direct C1 binding through specific sites in the transmembrane glycoprotein gp41. J Exp Med 174 1417–1424
Embretson J, Zupancic M, Ribas JL, Burker A, Rácz P, Tenner-Rácz K, Haase AT (1993) Massive overt infection of helper T lymphocytes and macrophages by HIV during the incubation period of AIDS. Nature 362: 359–362
Ennas MG, Chilosi M, Scarpa A, Lantini MS, Cadeddu G, Fiore-Donati L (1989) Isolation of multicellular complexes of follicular dendritic cells and lymphocytes: immunophenotypical characterization, electron microscopy and culture studies. Cell Tissue Res 257: 9–15
Fox CH, Cottler-Fox M (1992) The pathobiology of HIV infection. Immunol Today 13: 353–356
Freedman AS, Munro JM, Rice GE, Bevilacqua MP, Morimoto C, McIntyre BW, Rhynhart K, Pober JS, Nadler LM (1990) Adhesion of human B cells to germinal centers in vitro involves VLA-4 and INCAM-110. Science 249: 1030–1033

Freedman AS, Munro JM, Morimoto C, McIntyre BW, Rhynhart K, Lee N, Nadler LM (1992) Follicular non-Hodgkin's lymphoma cell adhesion to normal germinal centers and neoplastic follicles involves very late antigen-4 vascular cell adhesion molecule-1. Blood 79: 206–212

Garcia CF, Lifson JD, Engleman EG, Schmidt DM, Warnke RA, Wood GS (1986) The immunohistology of the persistent generalized lymphadenopathy syndrome (PGL). Am J Clin Pathol 86: 706–715

Gerdes J, Flad H-D (1992) Follicular dendritic cells and their role in HIV infection. Immunol Today 13: 81–83

Golding H, Robey FA, Gates FT III, Linder W, Beining PR, Hoffman T, Golding B (1988) Identification of homologous regions in human immunodeficiency virus I gp41 and human MHC class II β 1 domain. I. Monoclonal antibodies against the gp41-derived peptide and patient's sera react with native HLA class II antigens, suggesting a role for autoimmunity in the pathogenesis of acquired immune deficiency syndrome. J Exp Med 167: 914–923

Grimley PM, Davis GL, Kang Y-H, Dooley JS, Strohmaier J, Hoofnagle JH (1985) Tubuloreticular inclusions in peripheral blood mononuclear cells related to systemic therapy with α-interferon. Lab Invest 52: 638–649

Hardie DL, Johnson GD, Khan M, MacLennan IC (1993) Quantitative analysis of molecules which distinguish functional compartments within germinal centers. Eur J Immunol 23: 997–1004

Haynes BF, Robert-Guroff M, Metzgar RS, Franchini G, Kalyanaraman VS, Palker TJ, Gallo RC (1983) Monoclonal antibody against human T cell leukemia virus p19 defines a human thymic epithelial antigen acquired during ontogeny. J Exp Med 157: 907–920

Heinen E, Lilet-Leclercq C, Masson DY, Stein H, Boniver J, Kinet-Denoël C, Simar LJ (1984) Isolation of follicular dendritic cells from human tonsils and adenoids. II. Immunocytochemical characterization. Eur J Immunol 14: 267–273

Heinen E, Radoux D, Kinet-Denoël C, Moeremans M, De Mey J, Simar LJ (1985) Isolation of follicular dendritic cells from human tonsils and adenoids. III. Analysis of their Fc receptors. Immunology 54: 777–784

Heusermann U, Zurborn K-H, Schoeder L, Stutte HJ (1980) The origin of the dendritic reticulum cell: an experimental enzyme-histochemical and electron microscopic study on the rabbit spleen. Cell Tissue Res 209: 279–294

Homsy J, Meyer M, Tateno T, Clarkson S, Levy JA (1989) The Fc and not CD4 receptor mediates antibody enhancement of HIV infection in human cells. Science 244: 1357–1360

Humphrey JH, Greenan D, Sundaram V (1984) The origin of follicular dendritic cells in the mouse and the mechanism of trapping of immune complexes on them. Eur J Immunol 14: 859–864

Imazeki N, Senoo A , Fuse Y (1992) Is the follicular dendritic cell a primarily stationary cell? Immunology 76: 508–510

Janossy G, Pinching AJ, Bofill M, Weber J, McLaughlin JE, Ornstein M, Ivory K, Harris JR, Favrot M, Macdonald-Burns DC (1985) An immunohistological approach to persistent lymphadenopathy and its relevance to AIDS. Clin Exp Immunol 59: 257–266

Janossy G, Bofill M, Schuurman H-J (1991) Human B-lymphoid differentiation: normal versus malignant. Neth J Med 39: 232–243

Joling P, Van Wichen DF, Parmentier HK, Biberfeld P, Böttiger D, Tschopp J, Rademakers LHPM, Schuurman H-J (1992) Simian immunodeficiency virus (SIV_{sm}) infection of cynomolgus monkeys: effects on follicular dendritic cells in lymphoid tissue. AIDS Res Hum Retroviruses 8: 2021–2030

Joling P, Bakker LJ, Van Strijp JAG, Meerloo T, De Graaf L, Dekker MEM, Goudsmit J, Verhoef J, Schuurman H-J (1993) Binding of human immunodeficiency virus type-1 to follicular dendritic cells in vitro is complement dependent. J Immunol 150: 1065–1073

Joling P, Boom S, Johnson J, Dekker MEM, Van den Tweel JG, Schuurman H-J, Bloem AC (1994) Domain 5 of the intercellular adhesion molecule–1 (ICAM-1) is involved in adhesion of B-cells and follicular dendritic cells. In: In vivo immunology: Regulatory processes during lymphopoiesis and immunopoiesis. Heinen E, Defrense MP, Boniver J, Geenen V (eds) Plenum Publishing, London, pp 159–163

Jouault T, Chapuis F, Oliver R, Parravicini C, Bahraoui E, Gluckman JC (1989) HIV infection of monocytic cells: role of antibody-mediated virus binding to Fc–γ receptors. AIDS 3: 125–133

June RA, Schade SZ, Bankowski KJ, Kuhns M, McNamara A, Lint TF, Landay AL, Spear GT (1991) Complement and antibody mediate enhancement of HIV infection by increasing virus binding and provirus formation. AIDS 5: 269–274

Klatzmann D, Champagne E, Chamaret S, Gruest J, Guetard D, Hercend T, Gluckman JC, Montagnier L (1984) T lymphocyte T4 molecule behaves as the receptor for human retrovirus LAV. Nature 312: 767–768

Klaus GGB, Humphrey JH, Kunkl A, Dongworth DW (1980) The follicular dendritic cell: its role in antigen presentation in the generation of immunologic memory. Immunol Rev 53: 3–28
Koopman G, Pals ST (1992) Cellular interactions in the germinal center: role of adhesion receptors and significance for the pathogenesis of AIDS and malignant lymphoma. Immunol Rev 126: 21–45
Koopman G, Parmentier HK, Schuurman H-J, Newman W, Meijer CJLM, Pals ST (1991) Adhesion of human B cells to follicular dendritic cells involves both the lymphocyte function-associated antigen 1/intercellular adhesion molecule 1 and very late antigen 4/vascular cell adhesion molecule 1 pathways. J Exp Med 173: 1297–1304
Kosco MH, Gray D (1992) Signals involved in germinal center reactions. Immunol Rev 126: 63–76
Kosco MH, Pflugfelder E, Gray D (1992) Follicular dendritic cell-dependent adhesion and proliferation of B cells in vitro. J Immunol 148: 2331–2339
Kostianovsky M, Kang YH, Grimley PM (1983) Disseminated tubuloreticular inclusions in acquired immunodeficiency syndrome (AIDS). Ultrastruct Pathol 4: 331–336
Kozak C (1985) Retroviruses as chromosomal genes in the mouse. Adv Cancer Res 44: 295–336
Laman JD, Claassen E, Van Rooyen N, Boersma WJA (1989) Immune complexes on follicular dendritic cells as a target for cytolytic cells in AIDS. AIDS 3: 543–548
Langhoff E, Haseltine WA (1992) Infection of accessory dendritic cells by human immunodeficiency virus type 1. J Invest Dermatol 99: 89S–94S
Langhoff E, Terwilliger EF, Bos HJ, Kalland KH, Poznansky MC, Bacon OM, Haseltine WA (1991) Replication of human immunodeficiency virus type 1 in primary dendritic cell cultures. Proc Natl Acad Sci USA 88: 7998–8002
Langhoff E, Kalland KH, Haseltine WA (1993) Early molecular replication of human immunodeficiency virus type 1 in cultured blood-derived T-helper dendritic cells. J Clin Invest 91: 2721–2726
Larsson E, Kato N, Cohen M (1989) Human endogenous proviruses. In: Vogt PK (ed) Oncogenes and retroviruses. Springer, Berlin Heidelberg New York, pp 115–132 (Current topics in microbiology and immunology, Vol 148)
Le Tourneau A, Audouin J, Diebold J, Marche C, Tricottet V, Reynes M (1986) LAV-like viral particles in lymph node germinal centers in patients with the persistent lymphadenopathy syndrome and the acquired immunodeficiency syndrome-related complex: an ultrastructural study of 30 cases. Hum Pathol 17: 1047–1053
Lifson J, Coutré S, Huang E, Engleman E (1986) Role of envelope glycoprotein carbohydrate in human immunodeficiency virus (HIV) infectivity and virus-induced cell fusion. J Exp Med 164: 2101–2106
Lilet-Leclercq C, Radoux D, Heinen E, Kinet-Denoël C, Defraigne JO, Houben-Defresne MP, Simar LJ (1984) Isolation of follicular dendritic cells from human tonsils and adenoids. I. Procedures and morphological characterization. J Immunol Methods 66: 235–244
Lindhout E, Mevissen ML, Kwekkeboom J, Tager JM, De Groot C (1993) Direct evidence that human follicular dendritic cells (FDC) rescue germinal centre B cells from death by apoptosis. Clin Exp Immunol 91: 330–336
Louis E, Philippet B, Cardos B, Heinen E, Cormann N, Kinet-Denoël C, Braum M, Simar LJ (1989) Intercellular contacts between germinal center cells. Mechanisms of adhesion between lymphoid cells and follicular dendritic cells. Acta Otorhinolaryngol Belg 43: 297–320
Luu J, Bockus D, Remington F, Bean MA, Kammar SP (1989) Tubuloreticular structures and cylindrical confronting cisternae: a review. Hum Pathol 20: 617–627
Macatonia SE, Patterson S, Knight SC (1989) Suppression of immune responses by dendritic cells infected with HIV. Immunology 67: 285–289
Macatonia DE, Lau R, Patterson S, Pinching AJ, Knight SC (1990) Dendritic cell infection, depletion and dysfunction on HIV-infected individuals. Immunology 71: 38–45
MacLennan IC, Liu YJ, Johnson GD (1992) Maturation and dispersal of B-cell clones during T cell-dependent antibody responses. Immunol Rev 126: 143–161
Maeda M, Muro H, Shirasawa H (1988) C1q production and C1q-mediated immune complex retention in lymphoid follicles of rat spleen. Cell Tissue Res 254: 543–551
Maeda K, Burton GF, Padgett DA, Conrad DH, Huff TF, Masuda A, Szakal AK, Tew JG (1992) Murine follicular dendritic cells and low affinity Fc receptors for IgE (FcεRII). J Immunol 148: 2340–2347
Masson D, Tschopp J (1987) A family of serine esterases in lytic granules of cytolytic T lymphocytes. Cell 49: 679–685
Masuda A, Kasajima T, Mori N, Oka K (1989) Immunohistochemical study of low affinity Fc receptor for IgE on reactive and neoplastic follicles. Clin Immunol Immunopathol 53: 309–320
Montefiori DC, Robinson WE, Hirsch VM, Modliszewski A, Mitchell WM, Johnson PR (1990) Antibody-dependent enhancement of simian immunodeficiency virus (SIV) infection in vitro by plasma of SIV-infected rhesus macaques. J Virol 64: 113–119

Montefiori DC, Stewart K, Ahearn JM, Zhou J, Zhou J (1993) Complement-mediated binding of naturally glycosylated and glycosylation-modified human immunodeficiency virus type 1 to human CR2 (CD21). J Virol 67: 2699–2709

Naylor PH, Naylor CW, Badamchian M, Wada S, Goldstein AL, Wang S-S, Sun DK, Thornton AH, Sarin PS (1987) Human immunodeficiency virus contains an epitope immunoreactive with thymosin α_1 and the 30-amino acid synthetic p17 group-specific antigen peptide HGP-30. Proc Natl Acad Sci USA 84: 2951–2955

Nguyen TD, Scheving LA (1987) Thymosin α_1: amino acid homology with peptide T from the human immunodeficiency virus envelope. Biochem Biophys Res Commun 145: 884–887

O'Hara CJ, Groopman JE, Federman M (1988) The ultrastructural and immunohistochemical demonstration of viral particles in lymph nodes from human immunodeficiency virus-related and non-human immunodeficiency virus-related lymphadenopathy syndromes. Hum Pathol 19: 545–549

Onerheim RM, Wang N-S, Gilmore N, Jothy S (1984) Ultrastructural markers of lymph nodes in patients with acquired immunodeficiency syndrome and in homosexual males with unexplained persistent lymphadenopathy. Am J Clin Pathol 82: 280–288

Öst Å, Baroni CD, Biberfeld P, Diebold J, Moragas A, Noël H, Pallesen G, Rácz P, Schipper M, Tenner-Ráczk, Van den Tweel JG (1989) Lymphadenopathy in HIV infection: histologic classification and staging. APMIS Suppl 8: 7–15

Pantaleo G, Graziosi C, Demarest JF, Butini L, Montroni M, Fox CH, Orenstein JM, Kotler DP, Fauci AS (1993a) HIV infection is active and progressive in lymphoid tissue during the clinically latent stage of disease. Nature 362: 355–358

Panteleo G, Graziosi C, Fauci AS (1993b) The role of lymphoid organs in the pathogenesis of HIV infection. Semin Immunol 5: 157–163

Parmentier HK, Van Wichen D, Sie-Go DMDS, Goudsmit J, Borleffs JCC, Schuurman H-J (1990) HIV-1 infection and virus production in follicular dendritic cells in lymph nodes. A case report, with analysis of isolated follicular dendritic cells. Am J Pathol 137: 247–251

Parmentier HK, Van der Linden JA, Krijnen J, Van Wichen DF, Rademakers LHPM, Bloem AC, Schuurman H-J, (1991a) Human follicular dendritic cells: isolation and characteristics in situ and in suspension. Scand J Immunol 33: 441–452

Parmentier HK, Van Wichen DF, Peters PJ, Tschopp J, De Weger RA, Schuurman H-J (1991b) No histological evidence for cytotoxic T cells in destruction of lymph-node follicle centers after HIV infection. AIDS 5: 778–780

Parmentier HK, Van Wichen DF, Gmelig Meyling FHJ, Goudsmit J, Schuurman H-J (1992) Epitopes of human immunodeficiency virus regulatory proteins tat, nef and rev are expressed in normal human tissue. Am J Pathol 141: 1209–1216

Parravicini CL, Klatzmann D, Jaffray P, Costanzi G, Gluckman JC (1988) Monoclonal antibodies to the human immunodeficiency virus p18 protein cross-react with normal human tissues. AIDS 2: 171–177

Parravicini CL, Petren AL, Vago L, Costanzi G, Gluckman JC, Gallo RC, Biberfeld P (1989) HIV encephalopathy and lymphadenopathy: cells associated with viral antigens. APMIS Suppl 8: 33–39

Patterson S, Gross J, Bedford P, Knight SC (1991) Morphology and phenotype of dendritic cells from peripheral blood and their productive and non-productive infection with human immunodeficiency virus type 1. Immunology 72: 361–367

Pekovic DD, Gornitsky M, Ajdukovic D, Dupuy J-M, Chausseau J-P, Michaud J, Lapointe N, Gilmore N, Tsoukas C, Zwadlo G, Popovic M (1987) Pathogenicity of HIV in lymphatic organs of patients with AIDS. J Pathol 152: 31–35

Peters PJ, Borst J, Oorschot V, Fukuda M, Krähenbühl O, Tschopp J, Slot JW, Geuze HJ (1991) Cytotoxic T lymphocyte granules are secretory lysosomes, containing both perforin and granzymes. J Exp Med 173: 1099–1109

Petrasch S, Perez-Alvarez C, Schmitz J, Kosco M, Brittinger G (1990) Antigenic phenotyping of human follicular dendritic cells isolated from nonmalignant and malignant lymphatic tissue. Eur J Immunol 20: 1013–1018

Petrasch S, Kosco M, Schmitz J, Wacker HH, Brittinger G (1992) Follicular dendritic cells in non-Hodgkin lymphoma express adhesion molecules complementary to ligands on neoplastic B cells. Br J Haematol 82: 695–700

Piris MA, Rivas C, Morente M, Rubio C, Martin C, Olivia H (1987) Persistent and generalized lymphadenopathy: a lesion of follicular dendritic cells? An immunohistologic and ultrastructural study. Am J Clin Pathol 87: 716–724

Porwit A, Bottiger B, Pallesen G, Bodner A, Biberfeld P (1989) Follicular involution in HIV lymphadenopathy. A morphometric study. APMIS 97: 153–165

Prevot S, Fournier JG, Tardivel I, Audouin J, Diebold J (1989) Detection by in situ hybridization of HIV I RNA in spleens of HIV I sero-positive patients with thrombocytopenic purpura. Pathol Res Pract 185: 187–193
Pugliese O, Viora M, Camponeschi B, Cordiali Fei P, Caprilli F, Chersi A, Evangelista M, Di Massimo AM, Colizzi V (1992) A gp 120 HIV peptide with high similarity to HLA class II β chains enhances PPD-specific and autoreactive T cell activation. Clin Exp Immunol 90: 170–174
Purcell DFJ, Deacon NJ, McKenzie IFC (1989) The human nonlineage antigen CD46 (HuLY-M5) and primate retroviral gp70 molecules share protein-defined antigenic determinants. Immunol Cell Biol 67: 279–289
Query CC, Keene JD (1987) A human autoimmune protein associated with U1 RNA contains a region of homology that is crossreactive with p30 gag antigen. Cell 51: 211–220
Rácz P, Tenner-Rácz K, Kahl C, Feller AC, Kern P, Dietrich M (1986) The spectrum of morphologic changes of lymph nodes from patients with AIDS or AIDS -related complexes. Progr Allergy 37: 81–181
Rácz P, Tenner-Rácz K, Van Vloten F, Schmidt H, Dietrich H, Gluckman JC, Letvin NL, Janossy G (1990) Lymphatic tissue changes in AIDS and other retrovirus infection: tools and insights. Lymphology 23: 85–91
Rademakers LHPM (1992) Dark and light zones of germinal centers of the human tonsil: an ultrastructural study with emphasis on heterogeneity of follicular dendritic cells. Cell Tissue Res 269: 359–368
Rademakers LHPM, Schuurman H-J, De Frankrijker JF, Van Ooyen A (1992a) Cellular composition of germinal centers in lymph nodes after HIV-1 infection: evidence for an inadequate support of germinal center B lymphocytes by follicular dendritic cells. Clin Immunol Immunopathol 62: 148–159
Rademakers LHPM, Van Wichen D, De Weger RA (1992b) Immunohistochemical localization of intermediate filament and contractile proteins in the tonsil with reference to follicular dendritic cells. In: Imhof BA, Berrih-Aknin S, Ezine S (eds) Lymphatic tissues and in vivo immune responses. Dekker, New York, pp 721–725
Ree HJ, Khan AA, Elsakr M, Liau S, Teplitz C (1993) Intercellular adhesion molecule-1 (ICAM-1) staining of reactive and neoplastic follicles. ICAM-1 expression of neoplastic follicle differs from that of reactive germinal center and is independent of follicular dendritic cells. Cancer 71: 2817–2822
Reisinger EC, Vogetseder W, Berzow D, Köfler D, Bitterlich G, Lehr HA, Wachter H, Dierich MP (1990) Complement-mediated enhancement of HIV-1 infection of the monoblastoid cell line U937. AIDS 4: 961–965
Reynes M, Aubert JP, Cohen JHM, Audouin J, Tricottet V, Diebold J, Kazatchkine MD (1985) Human follicular dendritic cells express CR1, CR2 and CR3 complement receptor antigens. J Immunol 135: 2687–2694
Ringler DJ, Wyand MS, Walsh DG, MacKey JJ, Chalifoux LV , Popovic M, Minassian AA, Sehgal PK, Daniel MD, Desrosiers RC (1989) Cellular localization of simian immunodeficiency virus in lymphoid tissues. I. Immunohistochemistry and electron microscopy. Am J Pathol 134: 373–383
Ritter J, Sepetjan M, Monier JC (1987) Lack of reactivity of anti-human immunodeficiency virus (HIV) p17/18 antibodies against α1 thymosin and of anti-α1 thymosin monoclonal antibody against p17/18 protein. Immunol Lett 16: 97–100
Robinson WE, Montefiori DC, Mitchell WM (1988) Antibody-dependent enhancement of human immunodeficiency virus type 1 infection. Lancet 1: 790–794
Robinson WE, Montefiori DC, Gillespie DH, Mitchell WM (1989) Complement-mediated, antibody-dependent enhancement of HIV-1 infection in vitro is characterized by increased protein and RNA syntheses and infectious virus release. J Acquir Immune Defic Syndr 42: 33–42
Rodriguez ER, Nasim S, Hsia J, Sandin RL, Ferreira A, Hilliard BA, Ross AM, Garrett CT (1991) Cardiac myocytes and dendritic cells harbor human immunodeficiency virus in infected patients with and without cardiac dysfunction: detection by multiplex, nested, polymerase chain reaction in individually microdissected cells from right ventricular endomyocardial biopsy tissue. Am J Cardiol 68: 1511–1520
Rosenberg YJ, Zack PM, White BD, Papermaster SF, Elkins WR, Eddy GA, Lewis MG (1993) Decline in the $CD4^+$ lymphocyte population in the blood of SIV-infected macaques is not reflected in lymph nodes. AIDS Res Hum Retroviruses 9: 639–646
Ruco LP, Pomponi D, Pigott R, Gearing AJ, Baiocchini A, Baroni CD (1992) Expression and cell distribution of the intercellular adhesion molecule, vascular cell adhesion molecule, endothelial leukocyte adhesion molecule, and endothelial cell adhesion molecule (CD31) in reactive human lymph nodes and in Hodgkin's disease. Am J Pathol 140: 1337–1344
Sarin PS, Sun DK, Thornton AH, Naylor PH, Goldstein AL (1986) Neutralization of HTLV-III/LAV replication by antiserum to thymosin α_1. Science 232: 1135–1137

Schiaffino E, Bestetti-Bosisio M, Toia G, Onida L, Riboli P, Schmid C (1986) Ultrastructural alterations and virus-like particles in lymph nodes of drug addicts with lymphadenopathy syndrome (LAS). Pathol Res Pract 181: 755–760

Schnizlein CT, Kosco MH, Szakal AK, Tew JG (1985) Follicular dendritic cells in suspension: identification, enrichment, and initial characterization indicating immune complex trapping and lack of adherence and phagocytic activity. J Immunol 134: 1360–1368

Schmitz J, Petrasch S, Van Lunzen J, Rácz P, Kleine HD, Hufert F, Kern P, Schmitz H, Tenner-Rácz K (1993) Optimizing follicular dendritic cell isolation by discontinuous gradient centrifugation and use of the magnetic cell sorter (MACS). J Immunol Methods 159: 189–196

Schriever F, Nadler LM (1992) The central role of follicular dendritic cells in lymphoid tissues. Adv Immunol 51: 243–284

Schriever F, Freedman AS, Messner E, Lee G, Daley JS, Nadler LM (1989) Isolated human follicular dendritic cells display a unique antigenic phenotype. J Exp Med 169: 2043–2058

Schriever F, Freeman G, Nadler LM (1991) Follicular dendritic cells contain a unique gene repertoire demonstrated by single-cell polymerase chain reaction. Blood 77: 787–791

Schuurman HJ, Kluin PM, Gmelig-Meyling FHJ, Van Unnik JAM, Kater L (1985) Lymphocyte status of lymph node and blood in acquired immunodeficiency syndrome (AIDS) and AIDS-related complex disease. J Pathol 147: 269–280

Schuurman H-J, Krone WJA, Broekhuizen R, Goudsmit J (1988a) Expression of RNA and antigens of human immunodeficiency virus type -1 (HIV-1) in lymph nodes from HIV-infected individuals. Am J Pathol 133: 516–524

Schuurman H-J, Van Baarlen J, Krone WJA , Huber J (1988b) The thymus in the acquired immune deficiency syndrome. Thymus Update 1: 171–189

Schuurman H-J, Krone WJA, Broekhuizen R, Van Baarlen J, Van Veen P, Goldstein AL, Huber J, Goudsmit J (1989) The thymus in acquired immune deficiency syndrome. Comparison with other types of immunodeficiency diseases, and presence of components of human immunodeficiency virus type 1. Am J Pathol 134: 1329–1338

Schuurman H-J, Joling P, Van Wichen DF, Tobin D, Van der Putte SCJ (1993) Epitopes of human immunodeficiency virus (HIV-1) regulatory proteins tat, nef and rev are expressed in skin in atopic dermatitis. Int Arch Allergy Immunol 100: 107–114

Sellheyer K, Schwarting R, Stein H (1989) Isolation and antigenic profile of follicular dendritic cells. Clin Exp Immunol 78: 431–436

Sodroski J, Goh WC, Rosen C, Campbell K, Haseltine WA (1986) Role of the HTLV-III/LAV envelope in syncytium formation and cytopathicity. Nature 322: 470–474

Sölder BM, Schulz TF, Hengster P, Löwer J, Larcher C, Bitterlich G, Kurth R, Wachter H, Dierich MP (1989) HIV and HIV-infected cells differentially activate the human complement system independent of antibody. Immunol Lett 22: 135–145

Spear GT, Landay AL, Sullivan BL, Dittel B, Lint TF (1990) Activation of complement on the surface of cells infected by human immunodeficiency virus. J Immunol 144: 1490–1496

Spiegel H, Herbst H, Niedobitek G, Foss H-D, Stein H (1992) Follicular dendritic cells are a major reservoir for human immunodeficiency virus type 1 in lymphoid tissues facilitating infection of $CD4^+$ T-helper cells. Am J Pathol 140: 15–22

Stahmer I, Zimmer JP, Ernst M, Fenner T, Finnern R, Schmitz H, Flad H-D, Gerdes J (1991) Isolation of normal human follicular dendritic cells and CD4-independent in vitro infection by human immunodeficiency virus (HIV–1). Eur J Immunol 21: 1873–1878

Szakal AK, Kosco MH, Tew JG (1989) Microanatomy of lymphoid tissue during humoral immune responses: structure-function relationships. Annu Rev Immunol 7: 91–109

Szakal AK, Kapasi ZF, Masuda A, Tew JG (1992) Follicular dendritic cells in the alternative antigen transport pathway: microenvironment, cellular events, age and retrovirus related alterations. Semin Immunol 4: 257–265

Takeda A, Tuazon CU, Ennis FA (1988) Antibody-enhanced infection by HIV–1 via Fc-receptor-mediated entry. Science 242: 580–583

Tenner-Rácz K (1988) Human immunodeficiency virus-associated changes in germinal centers of lymph nodes and relevance to impaired B-cell function. Lymphology 21: 36–43

Tenner-Rácz K, Rácz P, Dietrich M, Kern P (1985) Altered follicular dendritic cells and virus-like particles in AIDS and AIDS-related lymphadenopathy. Lancet 1: 105–106

Tenner-Rácz K, Rácz P, Bofill M, Schultz-Meyer A, Dietrich M, Kern P, Weber J, Pinching AJ, Veronese-Dimarzo F, Popovic M, Klatzmann D, Gluckman JC, Janossy G (1986) HTLV-III/LAV viral antigens in lymph nodes from homosexual men with persistent generalized lymphadenopathy and AIDS. Am J Pathol 123: 9–15

Tenner-Rácz K, Rácz P, Gartner S, Ramsauer J, Dietrich M, Gluckman JC, Popovic M (1989) Ultrastructural analysis of germinal centers in lymph nodes of patients with HIV-1-induced persistent generalized lymphadenopathy: evidence of persistence of infection. Prog AIDS Pathol 1: 29–40

Tschachler E, Groh V, Popovic M, Mann DL, Konrad K, Safai B, Eron L, Veronese FM, Wolff K, Stingl G (1987) Epidermal Langerhans cells: a target for HTLV-III/LAV infection. J Invest Dermatol 88: 233–237

Tsunoda R, Nakayama M, Onozaki K, Heinen E, Cormann N, Kinet-Denoël C, Kojima M (1990) Isolation and long-term cultivation of human tonsil follicular dendritic cells. Virchows Arch [B] 59: 95–105

Tuijnman WB, Van Wichen DF, Schuurman H-J (1993) Tissue distribution of human IgG Fc receptors CD16, CD32 and CD64: an immunohistochemical study. APMIS 101: 319–329

Van den Berg TK, Dopp EA, Daha MR, Kraal G, Dijkstra CD (1992) Selective inhibition of immune complex trapping by follicular dendritic cells with monoclonal antibodies against rat C3. Eur J Immunol 22: 957–962

Vroom ThM, Van Wichen DF, Broekhuizen R, Van den Tweel JG, Joling P (1993) Adhesion molecules in lymphoid tissues demonstrated by immunohistochemistry. Tissue Antigens 42: 269

Wacker HH, Radzun HJ, Parwaresch MR (1990) Accessory cells in normal human and rodent lymph nodes: morphology, phenotype, and functional implications. In: Grundmann E, Vollmer E (eds) Reaction patterns of the lymph node. Springer, Berlin Heidelberg New York, pp 193–218 (Current topics in pathology, Vol 84)

Warner TFCS, Uno H, Gabel C, Tsai C-C (1984) a comparative ultrastructural study of virions in human pre-AIDS and simian AIDS. Ultrastructural Pathol 7: 251–259

Wilkinson DA, Freeman JD, Goodchild NL, Kelleher CA, Mager DL (1990) Autonomous expression of RTLV-H endogenous retrovirus-like elements in human cells. J Virol 64: 2157–2167

Wood GS (1990) The immunohistology of lymph nodes in HIV infection: a review. Prog AIDS Pathol 2: 25–32

Yamakawa M, Imai Y (1992) Complement activation in the follicular light zone of human lymphoid tissues. Immunology 76: 378–384

Yoffe B, Lewis DE, Petrie BL, Noonan CA, Melnick JL, Hollinger FB (1987) Fusion as a mediator of cytolysis in mixtures of uninfected CD4+ lymphocytes and cells infected by human immunodeficiency virus. Proc Natl Acad Sci USA 1987; 84: 1429–1433

Yoshida K, Van den Berg TK, Dijkstra CG (1994) The functional state of follicular dendritic cells in severe combined immunodeficient (SCID) mice: role of lymphocytes. Eur J Immunol 24: 464–468

Zwirner J, Felber E, Schmidt P, Riethmüller G, Feucht HE (1989) Complement activation in human lymphoid germinal centres. Immunology 66: 270–277

Follicular Dendritic Cells in Malignant Lymphomas

S. Petrasch

1 Introduction

The malignant lymphomas now known as Hodgkin's disease were first described by Thomas Hodgkin in 1832. This disease is characterized by the presence of large, multinucleate cells, called Reed-Sternberg cells, which are scattered irregularly throughout Hodgkin's infiltrates. Intermingled with the Reed-Sternberg cells is a mixed population of lymphocytes, histiocytes, eosinophils, plasma cells, and neutrophils. Histological subclassification of Hodgkin's disease relies principally on variations in the proportions of lymphocytes and histiocytes. Thus, the lymphoma is subdivided into four types: lymphocyte predominant including the nodular paragranuloma, nodular sclerosis, mixed cellularity, and lymphocyte depleted (Lukes and Butler 1966). An estimated 4 new patients with Hodgkin's

---

Medizinische Klinik, Ruhr-Universität Bochum, Knappschaftskrankenhaus, In der Schornau 23-25, 44892 Bochum, Germany disease per 100 000 population will be detected yearly. By chemotherapy and/or radiation approximately 75% of patients with Hodgkin's disease can be cured.

Malignant lymphomas other than Hodgkin's disease are called non-Hodgkin's lymphomas (NHL). In the Western world, they are derived in 90% of cases from B lymphocytes and in 10% from T cells or monocytes. More than 5 new NHL patients per 100 000 population per year are diagnosed. Various systems have evolved for the classification of NHL. While the Rappaport classification enjoys popularity mainly in the United States (RAPPAPORT et al. 1956), many European countries apply a scheme proposed by Lennert at the University of Kiel (LENNERT et al. 1975). He related morphology to lymphocyte lineage. The main subtypes of his classification are the lymphocytic, the centrocytic, and the centroblastic–centrocytic lymphoma (all of low-grade malignancy) and the centroblastic, immunoblastic, lymphoblastic, and Burkitt's lymphoma (all highly malignant neoplasias). The cytomorphological criteria of these entities will be explained in detail below. Only 25% of the patients suffering from NHL are cured from their disease.

Most research on malignant lymphoma has focused on the nature of Reed-Steenberg cells in Hodgkin's disease or neoplastic lymphocytes in NHL. However, information dealing with other components of lymphoma tissue such as cells of the mononuclear phagocytic system and the dendritic cell family may be as important. Within normal lymphatic tissue, follicular dendritic cells (FDC) form a web-like network throughout primary B cell follicles and germinal centers. Due to their long cytoplasmic processes, they are in close contact with many of the neighboring lymphocytes. FDC are also present in NHL derived from the follicular center or the mantle zone (STEIN et al. 1982a). In the centroblastic–centrocytic lymphoma, nodules resembling the secondary follicles within the nonmalignant lymphatic tissue are detected. These nodules contain FDC which constitute a well-defined, dense, spherical network. The monoclonal B cells with a centrocytic and centroblastic cytology are contained within the cytoplasmic processes of the FDC, where they undergo proliferation (MORI et al. 1988). FDC are not only restricted to NHL of the B cell type. They have also been discovered in T cell lymphoma, i.e., angioimmunoblastic lymphadenopathy (AILD). Furthermore, in Hodgkin's disease, the nodular paragranuloma with lymphocyte predominance displays FDC as a constitutive cell type.

Tumor cells from up to 85% of patients with follicular lymphoma and up to 30% of patients with large cell lymphoma contain the t(14; 18)(q32; q21) translocation. This rearrangement is the result of a breakpoint on chromosome 14 onto which a gene from chromosome 18 is joined called *bcl*-2 (B-cell leukemia/lymphoma-2). As a result of the translocation, *bcl*-2 is brought into the proximity of the immunoglobulin (Ig)H enhancer. Several investigators have found indications that the *bcl*-2 protein may act as a survival factor. When the *bcl*-2 gene was transfected into an interleukin (IL)-3-dependent cell line, the cells survived in a G_0 state but did not proliferate. They escaped from apoptosis, a process also called programmed cell death (VAUX et al. 1988). Interestingly, in vitro experiments have demonstrated that FDC are able to rescue germinal center B cells from entering apoptosis (PETRASCH et al. 1991; LINDHOUT et al. 1993). However, no experimental evidence for a contribution of FDC in the oncogenesis of malignant lymphoma has been published so far. On the other hand, several investigators were able to show that FDC stimulate the proliferation of non-neoplastic and neoplastic lymphocytes (SCHNITZLEIN et al. 1984; PETRASCH et al. 1992a).

This chapter will first focus on neoplasias with FDC involvement. The histological distribution of FDC within the different lymphoma entities and within FDC sarcoma will be described in detail. Subsequently, the antigenic phenotype of FDC in normal conditions and in neoplastic disorders will be summarized, paying special attention to the adhesion molecules expressed on FDC and neoplastic lymphocytes. Finally, the interactions between FDC, germinal center B lymphocytes, and lymphoma cells will be discussed.

2 Histological Distributcomn of Follicular Dendritic Cells

The detection of FDC by conventional light microscopy or histochemical techniques in human lymph nodes is difficult. The production of monoclonal antibodies (mAb) has enabled pathologists to study the distribution pattern of FDC in the lymphatic tissue. Of particular interest are those mAb specific for FDC, namely Ki-M4 (PARWARESCH et al. 1983) and R4/23 (NAIEM et al. 1983). The mAb Ki-M4P and Ki-FDC1 are adequate for formalin-fixed and paraffin-embedded tissue sections. They are well-established immunoreagents for the recognition of FDC in routine diagnostic and research work in human lymph nodes (TABRIZICHI et al. 1990) and in rats (WACKER et al. 1987), respectively.

2.1 Normal Lymphoid Tissue

FDC are restricted to the primary follicles and germinal centers of the secondary lymphoid tissue (NOSSAL et al. 1968; STEIN et al. 1982a). They may also be visualized in lymph follicles within various extralymphatic organs when chronic inflammations are accompanied by an activation of B cells, i.e., rheumatoid arthritis. In addition, FDC are detectable in the mantle zone that surrounds the germinal center. In the mantle zone their fusiform processes are more loosely arranged (WOOD et al. 1985).

Within the germinal center a dark zone rich in centroblasts and a light zone consisting predominantly of centrocytes can be distinguished. In the latter, the dendritic processes are in close association with the lymphocytes, while in the dark zone, the immunohistochemical staining of FDC is less intense. By electron microscopy, Rademakers observed undifferentiated FDC with only few cell organelles in the dark zone of the secondary follicles, while FDC in the light zone presented numerous mitochondria and their cytoplasmic processes carried electron-dense deposits (RADEMAKERS 1992).

FDC have not been identified in the extrafollicular regions of the lymphatic tissue, nor have they been detected in specimen of normal thymus, liver, skin, or kidney (PARWARESCH et al. 1983.)

2.2 Hodgkin's Disease

In Hodgkin's disease, two histological patterns of FDC can be distinguished: FDC may form an expanded and disrupted network occupied by nongerminal center cells. The follicle-occupying cells are epitheloid, lymphatic, and histiocytic cells (L and H). This pattern is found in 90% of the cases of nodular paragranuloma with lymphocyte predominance and in 50% of the cases of nodular sclerosis (ALVAIKKO et al. 1991). Furthermore, in some patients with Hodgkin's disease presenting a mixed cellularity, a network-like distribution of FDC has been observed, with the follicle-occupying cells being Sternberg-Reed, Hodgkin's, and epitheloid cells. In all the other cases, including the lymphocyte-depleted subtype, FDC are either rare or absent. In immunostained cryostate sections, only remnants of FDC processes are visible in these cases. Thus, the disappearance of lymphocytes and the abundance of sclerosis in Hodgkin's disease is accompanied by a loss of FDC (ALVAIKKO et al. 1991). Interestingly, these latter entities are associated with a poor prognosis.

2.3 Non-Hodgkin's Lymphoma

In NHL derived from germinal center cells, i.e., centroblastic–centrocytic or centroblastic lymphoma, nodules resembling the secondary follicles within nonmalignant lymphoid tissues are detected. The neoplastic centroblasts and centrocytes are intermingled with the FDC. Furthermore, FDC may be present in non-germinal-center cell derived B cell NHL, i.e., centrocytic and lymphocytic lymphoma. Finally, FDC are frequently detected in low-grade B cell lymphoma of the gastrointestinal tract and in some cases of T cell neoplasias, i.e., (AILD).

2.3.1 Germinal Center Cell-Derived Non-Hodgkin's Lymphoma

The centroblastic–centrocytic lymphoma and the centroblastic lymphoma of the Kiel classification are termed "malignant follicular lymphoma" and "diffuse malignant lymphoma", respectively, according to the Working Formulation. The neoplastic nodules of the centroblastic–centrocytic lymphoma consistently display FDC, constituting a well-defined, dense, spherical meshwork. In some cases, however, the number of FDC is reduced and the demarcation of the network is not as clear-cut as it is in reactive follicles. In this entity, the monoclonal B lymphocytes, which often have strongly indented nuclei, are enwrapped tightly by the cytoplasmic processes of the FDC (STEIN et al. 1984). Although in some cases of centroblastic–centrocytic lymphoma the follicle mantle still exists, a subdivision of the pseudofollicles into dark and light zones is not detected. The number of proliferating cells within the neoplastic germinal centers is lower than in reactive germinal centers, and starry sky macrophages may be absent (LENNERT and FELLER 1990).

FDC have been identified only in some cases of highly malignant NHL derived from germinal center cells, i.e., centroblastic lymphoma. In this subtype, FDC are less densely packed or are present in irregular clusters or clumps, sometimes showing an asymmetrical layout (STEIN et al. 1984).

2.3.2 Non-Germinal-Center Cell-Derived Non-Hodgkin's Lymphoma

In the centrocytic lymphoma, also called "mantle zone lymphoma" by WEISENBURGER et al. (1982), FDC are consistently present. In this disease, the neoplastic B cells are small to medium in size, displaying only barely detectable cytoplasma. Their nuclei are usually indented, containing a fine chromatin structure. Congruent with the non-nodular growth pattern revealed in the histological examination of centrocytic lymphoma (LENNERT et al. 1978), a diffuse distribution of the FDC can be seen by microscopy. In the centrocytic lymphoma, the cytoplasmic extensions of the FDC are rather round and short, presenting only vestigial ramifications. Interestingly, patients with centrocytic lymphoma and a weak positivity for the FDC-selective mAb Ki-M4 have a poor prognosis (STEIN et al. 1982a). Although the FDC are not well differentiated in this disease, they are closely associated with the neoplastic centrocytes. In some cases of centrocytic lymphoma, reactive secondary follicles with a well-established web of FDC may still be preserved.

FDC are also found in non-follicular-center cell lymphoma, when a pseudonodular growth pattern is observed. By immunohistochemistry with the mAb RFD-3, CHILOSI et al. (1985) detected FDC in 50% of the bone marrow samples from patients with a nodular marrow involvement of B cell chronic lymphocytic leukemia. Furthermore, FDC infiltrating lymphoma tissue have been observed in some patients with Burkitt's lymphoma (STEIN et al. 1984). This disease, frequently associated with a 8/14 translocation, is endemic in Africa and obviously arises in a state of Epstein-Barr virus-induced polyclonal B cell proliferation (DALLA FAVERA et al. 1982). Other forms of lymphoblastic lymphoma, even those with a nodular growth pattern, are devoid of FDC (MAEDA et al. 1993a). Furthermore, no FDC involvement has been found in prolymphocytic leukaemia, hairy cell leukemia, or multiple myeloma.

2.3.3 Lymphoma of Mucosa-Associated Lymphoid Tissue

Malignant lymphoma arising from mucosa-associated lymphoid tissue (MALT) are also called MALT lymphoma. Their oncogenesis is not yet fully understood; however, their is evidence for their derivation from the B cell native to mucosa. PARSONNET et al. (1994) were able to show that MALT lymphoma affecting the stomach is associated with a previous *Helicobacter pylorus* infection. Immunohistochemical analysis revealed the relationship between MALT lymphoma and the B lymphocytes of the marginal zone surrounding reactive follicles (SPENCER et al. 1985). MALT lymphomas are composed of plasma cells, lymphoplasmocytoid cells, and centrocytoid cells. These cellular components may derive from one tumor cell clone but arrest at different states of maturation. In this lymphoma type, the follicular mantle is infiltrated by lymphoma cells, while regressive changes occur in the germinal centers. The epithelium of the mucosa is infiltrated by centrocytoid cells, termed "lympho-epithelial lesion" by ISAACSON and WRIGHT (1984). In MALT lymphoma, an increase in FDC can be consistently detected (TIEMANN und PARWARESCH 1991). TABRIZCHI et al. (1990) surveyed 97 cases of low-grade B cell lymphoma of the MALT type. They observed small clusters of FDC, randomly distributed throughout the tumor, and interpreted them as tumor-associated abortive follicles. These abortive follicles, consisting of de novo developed convolutions of FDC, can only be detected by immunohistochemical methods, and not by conventional morphology. In other cases and similar to centrocytic lymphoma, a diffuse increase of FDC may be found in lymphoma of the MALT type (PETRASCH et al. 1994).

2.3.4 Angioimmunoblastic Lymphadenopathy

AILD, a subtype of peripheral T cell lymphomas, show a diffuse effacement of an arborizing proliferation of postcapillary high endothelial venules. These high endothelial venules are placed ontop of periodic acid-Schiff (PAS)-positive basement membranes. Clusters of large T cells with clear cytoplasma are intermingled with small to intermediate-sized lymphocytes, eosinophils, and plasma cells. Figure 1 shows the presence of FDC in AILD. FDC accumulate in bizzare-shaped aggregations, spreading out in large irregular sheets or burned-out germinal centers (LENNERT et al. 1979). LEUNG and coworkers (1993) stained peripheral T cell lymphomas with the FDC-selective mAb DRC-1. Only in AILD were FDC found, forming an expanded network of cells exceeding the confines of germinal centers. All other T cell neoplasias had either no FDC at all or FDC restricted to remnant follicle centers. We therefore recommend the routine staining of FDC in typing T cell lymphomas.

2.4 Follicular Dendritic Cell Sarcoma

In FDC sarcoma, the tumor cells are large with ill-defined cellular borders and irregular nuclei. The malignant cells are mixed with a polyclonal population of lymphoid cells. The neoplastic FDC may be bi- or polynucleated. By microscopical analysis, a storiform-like growth pattern is revealed (PALLESEN and MYHRE-JENSEN 1987). In the report by PALLESEN and MYHRE-JENSEN, evidence for the FDC-derived neoplasias was given by the immunocytochemical staining pattern, identical to that of FDC in the normal lymphatic tissue. Thus, tumor cells were positive for the FDC-selective mAb and, in addition, they expressed CD21 (EBV receptor), CD23 (low-affinity Fc-IgE receptor), and Cd35 (C3b receptor). CHAN and coworkers (1994) observed a FDC tumor of the oral cavity. The neoplasia was characterized

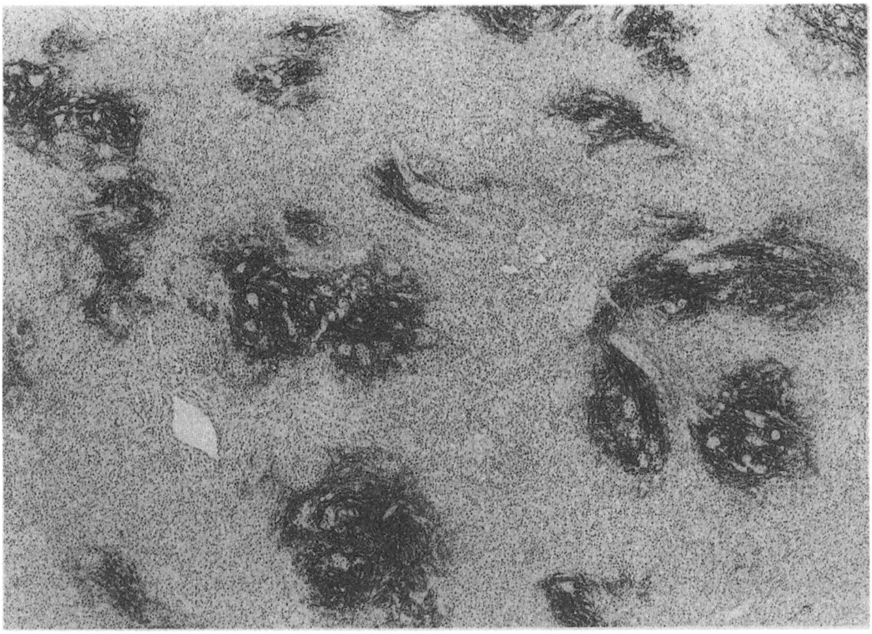

Fig. 1. Angioimmunoblastic lymphoma (AILD) displaying follicular dendritic cells in large, irregular sheets (*black areas*) engulfing high endothelial venules (paraffin-embedded lymph node, immunostained with the monoclonal antibody Ki-M4, selective for follicular dendritic cells)

by sheets, whorls, and storiform arrays of spindly and syncytial-appearing cells with oval nuclei with a delicate chromatin structure.

2.5 Extranodal Lymphoma

Malignant lymphoma detected in biopsy samples of liver, skin, or bone marrow are called extranodal because of their localization outside the lymphatic tissue. In 1983, NAIEM et al. observed a patient with a follicular lymphoma infiltrating the kidney. Within the neoplastic nodules, a dense, red mass of FDC was detected by immunohistochemistry. FDC are most predominant in extranodal lymphomas of the thyroid (FELLBAUM et al. 1993). Thyroid lymphomas show the morphological features of MALT-type lymphomas. Interestingly, the thyroid lymphomas examined by FELLBAUM et al. were highly malignant in 18 of 19 cases. These authors also detected FDC in some cases of malignant lymphoma of the kidney and of the testis. The detection of extranodal NHL with FDC involvement suggests that FDC immigrate into, or are newly formed within, the neoplastic microenvironment.

3 Antigen Profile of Follicular Dendritic Cells

Because of the close association between FDC and the surrounding lymphocytes, expression of surface molecules on FDC cannot be easily assessed when cryostat sections of lymph nodes are examined. Furthermore, the immunophenotype of human FDC varies according to their topographic localization within the follicles, indicating the heterogeneity of FDC. This may explain why the analysis of the antigenic profile of isolated FDC has also resulted in controversial observations.

3.1 Nonmalignant Tissue

Our own data indicate that FDC express molecules also present on B lymphocytes, namely CD20 (Bp35), CD21 (EBV receptor), and CD24 (Phosphatidylinositol-linked glycoprotein), while they are negative for anti-CD19 (Pan-B) and anti-CD 22 (Bgp135) (Petrasch et al. 1989. 1990). Other investigators, evaluating the immunophenotype of FDC in cryostat sections, have attributed the latter two molecules to FDC (Stein et al. 1982b; Johnson et al. 1986b). As already pointed out, the immunophenotype of FDC varies according to their localization within the reactive follicles. The FDC population of the central portion of the germinal center displays CD9 (p24) and CD14 (gp 55), while the accessory cells visualized in the mantle zone of the follicle are negative for anti-CD9 and anti-CD14 (Carbone et al. 1988). FDC in the light zone present ramificated cytoplasmic processes with electron-dense deposits, indicating a high degree of differentiation (Rademakers 1992). Only these FDC express the low-affinity receptor for IgE (CD23), a molecule associated with a state of activation (Stein et al. 1980; Johnson et al. 1986a). According to the majority of investigators, FDC isolated from non-malignant lymphatic tissue carry receptors for C3bi (CD 11b), C3d (CD21), and C3b (CD35), components of the complement system (Schriever et al. 1989; Sellheyer et al. 1989; Petrasch et al. 1989). Membrane-bound immunoglobulin heavy chains μ and γ as well as both κ and λ light chains are detectable on the surface of FDC. Congruent with the finding of electron-dense deposits only associated with FDC in the light zone of secondary follicles, anti-immunoglobulin mAb react with no more than 50% of FDC in single cell suspensions (Petrasch et al. 1990). FDC do not stain for IgD or IgA expression.

Other molecules expressed on FDC include class II antigens (HLA-DR), CD45 (common leukocyte antigen), and the myelomonocytic marker CD14 (gp55) (Wood et al. 1985; Petrasch et al. 1990). The presence of these three surface proteins found predominantly on cells of the leukocyte lineage lends support to the hypothesis that FDC may be bone marrow derived. Furthermore, the glycoprotein gp50 (CD40), which is capable of giving a strong cell cycle progression signal and promoting homotypic adhesion has been detected on the surface of the immunoaccessory cells (Schriever et al. 1989).

FDC do not express the T lymphocyte marker CD1, (T6), CD3 (CD3 complex), CD4 (gp59), CD5 (Tp67), or CD8 (T8) (SCHRIEVER et al. 1989; SELLHEYER et al. 1989; PETRASCH et al. 1989).

3.2 Malignant Tissue

As with normal lymphoid tissue, heterogeneous FDC subpopulations have been described in NHL. Immunohistochemical analysis of mantle zone lymphomas, such as the centrocytic lymphoma, revealed that FDC do not express CD9 (p24) and CD14 (gp55), molecules visualized on FDC within the central portion of reactive follicles. However, anti-CD9 and anti-CD14 mAb stained with FDC in follicular center cell-derived NHL (CARBONE et al. 1987). Thus the distribution of CD9 and CD14-negative/positive FDC within the normal lymphatic tissue is maintained in neoplasias related to these localizations. These findings are in line with an investigation of IMAI et al. (1990), who described a loss of FDC membrane antigens in lymphomas with a diffuse growth pattern, as compared to FDC in follicular lymphoma.

An antigenic phenotyping of FDC isolated from NHL was performed by our group (PETRASCH et al. 1990). The immunophenotype resembled FDC isolated from hyperplastic tonsils, suggesting that the immunological capacity, i.e. antigen trapping of FDC, is not compromised by the neoplastic condition. However, the amount of immunoglobulin deposits on FDC in NHL seems to decrease along with the neoplastic transformation, and some investigators obtained negative results when labeling lymphoma-associated FDC for surface immune complexes (BRAYLAN and RAPPAPORT 1973). MAEDA et al. (1993b) visualized focal or reticular depositions of immunoglobulins and complement components within neoplastic follicles. These authors also suggest that FDC in NHL still have the functional ability to trap and retain antigen. Although the immunological properties of FDC, i.e., immune complex trapping, might not be compromised by the neoplastic conditions, it can be hypothized that the lack of surface immunoglobulins on the FDC membrane in some NHL is due to a hindered processing of antigen and impeded formation of immune complexes within the lymphoma tissue.

3.3 Adhesion Molecules on Follicular Dendritic Cells in Malignant Lymphomas

Adhesion molecules are involved in cell migration, homing, localization, and retention of cells in certain microenvironments. When FDC and lymphocytes from NHL are incubated together in cell cultures, they spontaneously form small cellular aggregates with one FDC enwrapping several B cells. FREEDMAN et al. (1990) and KOOPMANN and coworkers (1991) were able to demonstrate that with cells from nonmalignant tissue, this adhesion is mediated by the lymphocyte function-associated antigen-1α (LFA-1α/CD11a), LFA-1β (CD18), the very late antigen-4 (VLA-4/CD49d), and the intracellular adhesion molecule-1 (ICAM-1/CD54) on B cells, and by ICAM-1 (CD54), the vascular cell adhesion molecule-1 (VCAM-1), and the complement receptor C3bi (Cd11b) on FDC. FDC isolated from lymph nodes of patients with centroblastic–centrocytic lymphoma are also positive for ICAM-1 and C3bi (Petrasch et al. 1992 b). Furthermore, Freedman and coworkers (1992) observed that follicular NHL cells attach to neoplastic follicles in vitro. This adhesion was inhibited by the addition of mAb directed against VLA-4 and VCAM-1. These data indicate that within lymphoma tissue a normal adhesive interaction occurs which accounts for the localization of neoplastic B cells in the proximity of FDC.

Our group analyzed the adhesion molecules on neoplastic B cells in lymph nodes and in the peripheral blood. B cells within the neoplastic tissue express CD11a, CD18, and CD54, while neoplastic lymphocytes in the peripheral blood of patients with a leukemic course of centroblastic–centrocytic lymphoma have lost all or part of these molecules (Petrasch et al. 1992b). In line with these data is the finding by Stauder and coworkers (1989), who reported that the loss of ICAM-1 (CD54) in low-grade NHL correlates with the presence of a leukemic phase. Thus, neoplastic centrocytes and centroblasts, due to their lack of CD11a, CD18, and CD54 surface molecules may detach from FDC, thus leaving the lymph node and invading new compartments. According to Freedman (1993), this aberrant expression and/or function of adhesion receptors provides an explanation for the biological and clinical behaviour of NHL.

In reactive B cell follicles, components of the extracellular matrix such as fibrils of collagen and fibrinogen contribute to the peculiar microarchitecture of the nodular structures. FDC isolated from NHL express adhesion receptors complementary to those on extracellular matrixes. Thus, CD51 (VNR α-chain), a ligand for fibrinogen, and CD49c (VLA α_3-chain), a ligand for collagen, were identified on the surface of FDC isolated from malignant lymphomas (Petrasch et al. 1992b). The adhesion between these extracellular components and the cytoplasmic processes of FDC may play a crucial role in the constitution and integrity of the neoplastic nodules in NHL. The antigenic profile of FDC is summarized in Table 1.

4 Interactions Between Follicular Dendritic Cells and Non-Hodgkin's Lymphoma Lymphocytes

In the preceeding chapters, the enhancing effect of FDC on the activation and proliferation of germinal centre B cells in the normal lymphatic tissue has been described in detail. Mori and coworkers (1988) stained lymphoma cryostat sections with the mAb Ki-67, a reagent specific for a nuclear antigen only present in the late G_1 to M phase. They visualized the majority of proliferating lymphocytes localized in the proximity of FDC. In low-grade NHL, only a small proportion of

Table 1. Antigen profile of follicular dendritic cells (FDC) from non-Hodgkins lymphoma (NHL) lymph nodes

Molecule	Name	Reactivity
CD1	T6	–
CD3	CD3 complex	–
CD4	gp59	–
CD8	T8	–
CD9	p24	+
CD10	CALLA	–
CD11a	Leukocyte function antigen-1	–
CD11b	C3bi receptor	+
CD13	Aminopeptidase N	–
CD14	gp55	+
CD16	Low affinity Fc receptor for IgG	+
CD18	LFA-1	–
CD19	Pan-B	–
CD20	Bp35	+
CD21	EBV receptor	+
CD22	BgP 135	–
CD23	Low affinity Fc-IgE receptor	+/–
CD24	PI-linked glycoprotein	+/–
CD29	Integrin β_1-chain	+/–
CD32	gp40	+
CD35	C3b receptor	+
CD38	T10	+
CD40	gp50	+
CD45	T200	+/–
CD49b	VLA α_2-chain	+/–
CD49c	VLA α_3-chain	+/–
CD49e	VLA α_5-chain	+
CD49f	VLA α_6-chain	+
CD51	VNR α-chain	+/–
CD54	Intercellular adhesion molecule-1	+
CD58	Leukocyte function antigen-3	–
HLA-DR	Human leukocyte alloantigen	+
α-chain	Immunoglobulin α-chain	–
γ-chain	Immunoglobulin γ-chain	+/–
δ-chain	Immunoglobulin δ-chain	–
κ-chain	Immunoglobulin κ-chain	+/–
λ-chain	Immunoglobulin λ-chain	+/–
μ-chain	Immunoglobulin μ-chain	+/–
E/L/P selectins	Selectins	–
VCAM-1	Vascular cell adhesion molecule-1	+/–

FDC were isolated from lymph nodes of patients with NHL. Expression of surface antigens was determined by immunocytochemistry.
+, positive, +/–, only part of the FDC we positive; –, negative; Ig, immunoglobulin; LFA, lymphocyte function-associated antigen; EBV, Epstein-Barr virus; PI, phosphatidylinositol; VLA, very late antigen; CALLA, common acute lymphocytic leukemia antigen.

lymphocytes proliferate, while in highly malignant lymphomas the Ki-67-positive fraction amounts to up to 58% (Schwartz et al. 1989). Frequently, FDC form a pseudonodular-like pattern in diffuse-type, low-grade NHL. These nodules contain dividing lymphocytes, thus forming centers of proliferation (Chilosi et al. 1985; Ratech et al. 1988).

In vitro studies with cultured FDC isolated from NHL have not been reported until recently. FDC isolation from lymphoma tissue has been hindered by the close association between FDC and the neoplastic lymphocytes. Furthermore, only a few lymph nodes of patients with NHL are available for in vitro assays. In our laboratory, we applied an enzyme cocktail to digest lymphoma tissue. Subsequently, the obtained single cell suspensions were layered on top of a bovine serum albumin gradient and centrifuged at 8500 *g*. Following isolation and enrichment of FDC, the spontaneous formation of cell aggregates with lymphoma lymphocytes was observed. Each FDC-dependent cellular cluster contained one to nine lymphoma cells. The expression of Ki-67 by the NHL cells in FDC-associated clusters was then evaluated. A considerable number of lymphocytes enclosed by the FDC processes were in late G_1 to M phase of the cell cycle. Furthermore, the number of cells outside the clusters staining positive for Ki-67 was significantly lower as compared to the neoplastic lymphocytes involved in cluster formation. These data suggested that FDC provide signals leading to the continued stimulation of NHL B cells. In a second investigation, we added [^{3}H] thymidine to the FDC/NHL B lymphocyte cell cultures. We were thus able to demonstrate that in cell preparations from lymphoma tissue enriched with FDC, an increase in the uptake of [^{3}H] thymidine occurs, even in the absence of mitogens. In our system, the distribution of [^{3}H] thymidine-positive cells was not limited to the cellular aggregates. However, at the end of the 24-h culture period, the ratio of [^{3}H] thymidine-positive cells inside versus outside the FDC-dependent clusters amounted to 6.7:1 and increased to 16.8:1 after 72 h of incubation. These observations suggest that in NHL, as in the normal tissue, the FDC cell surface may provide stimulatory signals for neoplastic lymphocytes (Petrasch et al. 1992a).

Different reagents have been proposed to stimulate the proliferation of NHL lymphocytes. In the presence of high molecular weight B cell growth factor, neoplastic B cells show an extensive proliferation (Alvarez-Mon et al. 1989). Chronic lymphocytic leukemia (CLL) B cells can be released from their maturation block by interleukin-2 and anti-Ig reagents or by interleukin-4 and immobilized anti-CD40 mAb (De France et al. 1991). It has already been mentioned that FDC within the neoplastic condition express CD40 on their surface and carry membrane-bound immunoglobulins.

Follicular B lymphocytes with a rearranged *Bcl-2* gene escape from apoptosis (Vaux et al. 1988). Avoiding apoptosis is considered to be an important step in the oncogenesis of germinal center cell-derived malignant lymphoma. When germinal center B cells isolated from human tonsils are put in culture, they quickly die by apoptosis. When immunoglobulin-coated sheep red blood cells or soluble anti-CD40 is added to these cultures, the germinal center B cells are rescued from death by apoptosis (Liu et al. 1989). As described by several investigators, a direct membrane contact to FDC also rescues cultured B lymphocytes from apoptotic cell death (Petrasch et al. 1991; Lindhout et al. 1993). When the formation of FDC/B cell clusters is prevented by the addition of anti LFA-1 and anti VLA-4 antibodies to the single cell preparation, virtually all cultured lymphocytes die by apoptosis (Lindhout et al. 1993). It is tempting to hypothize that, via the immune complexes retained on their surface and via their membrane-bound CD40 molecules, FDC provide signals preventing the apoptosis of germinal center B lymphocytes. How this occurs is still undefined. Further investigations should focus on the role of these interactions in the oncogenesis of malignant lymphoma.

References

Alavaikko MJ, Hansmann M-L, Nebendahl C, Parwaresch MR, Lennert K (1991) Follicular dendritic cells in Hodgkin's disease. Am J Clin Pathol 95: 194–200

Alvarez-Moon M, De la Hera A, Caspar ML, Orfoa A, Casa J, Jordia J, Durantes A (1989) Proliferation of B-cells from chronic lymphocytic leukemia is selectively promoted by B-cell growth factor. Acta Haematol 81: 91–97

Braylan RC, Rappaport H (1973) Tissue immunoglobulins in nodular lymphomas as compared with reactive follicular hyperplasias. Blood 42: 579–589

Carbone A, Poletti A, Manconi R, Gloghini A, Volpe R (1987) Heterogenous in situ immunophenotyping of follicular dendritic cells in malignant lymphomas of B-cell origin. Cancer 60: 2919–2926

Carbone A, Manconi R, Poletti A, Volpe R (1988) Heterogeneous immunostaining patterns of follicular dendritic reticulum cells in human lymphoid tissue with selected antibodies reactive with different cell lineages. Hum Pathol 19: 51–56

Chan JK, Tsang WY, Ng CS, Tang SK, Yu HC, Lee AW (1994) Follicular dendritic cell tumors of the oral cavity. Am J Surg Pathol 18: 148–157

Chilosi M, Pizzolo G, Caligaris-Cappio F, Ambrosetti A, Vinante F, Morittu L, Bonetti F, Fiore-Donati L, Janossi G (1985) Immunohistochemical demonstration of follicular dendritic cells in bone marrow involvement of B-cell chronic lymphocytic leucaemia. Cancer 56: 328–332

Dalla Favera R et al (1982) Human c-myc oncogene is located on the region of chromosome 8 that is translocated in Burkitt lymphoma cells. Proc Natl Avad Sci USA 79: 7924

De France T, Fluckiger AC, Rousset F, Banchereau J (1991) In vitro activation of B-cells. 5th international workshop on CLL, Sitges, Barcelona, 26-28 April, p 13

Fellbaum C, Sträter J, Hansmann M-L (1993) Follicular dendritic cells in extranodal non-Hodgkin lymphomas of MALT and non-MALT type. Virchows ARCH [A] 423: 335–341

Freedman AS (1993) Expression and function of adhesion receptors on normal B cells and B cell non-Hodgkin's lymphomas. Semin Hematol 30: 318–328

Freedman AS, Munro JM, Rice GE, Bevilacqua MP, Morimoto C, McIntyre BW, Rhynhart K, Pober JS, Nadler LM (1990) Adhesion of human B cells to germinal centers in vitro involves VLA-4 and INCAM-110. Science 249: 1030–1033

Freedman AS, Munro JM, Morimoto C, McIntyre BW, Rhynhart K, Lee L, Nadler LM (1992) Follicular non-Hodgkin's lymphoma cell adhesion to normal germinal centers and neoplastic follicles involves very late antigen-4 and vascular cell adhesion molecule-1. Blood 79: 206–212

Hodgkin T (1832) On some morbid appearances of the absorbent glands and spleen. Med Chir Trans 17: 68

Imai Y, Matsuda M, Maeda K, Narabayashi M, Masunaga A (1990) Follicular dendritic cells in lymphoid malignancies-morphology, distribution, and function. In: Hanaoka M, Kadin ME, Mikata A, Watanabe S (eds) Lymphoid malignancy. Field and Wood, New York, pp 229–242

Isaacson PG, Wright DH (1984) Extranodal malignant lymphoma arising from mucosa-associated lymphoid tissue. Cancer 53: 2515–2524

Johnson GD, MacLennan ICM, Ling NR, Hardie DL, Nathan PD, Walker L (1986a) Reactivity of monoclonal antibodies of the B and L series with follicular dendritic cells in tissue sections and with lymphoblastoid cell lines. In: Reinherz EL, Haynes BF, Nadler LM, Bernstein ID (eds) Leukocyte typing II. Human B lymphocytes. Spinger, Berlin Heidelberg New York, pp 289–297

Johnson GD, Hardie DL, Ling NR, MacLennan ICM (1986b) Human follicular dendritic cells (FDC): a study with monoclonal antibodies (MoAb). Clin Exp Immunol 64: 205–213

Koopman G, Partmentier HK, Schuurman H-J, Newman W, Meijer CJLM Pals ST (1991) Adhesion of human B cells to follicular dendritic cells involves both the lymphocytes function-associated antigen 1/intercellular adhesion molecule 1 and very late antigen 4/vascular cell adhesion molecule 1 pathways. J Exp Med 173: 1297–1304

Lennert K, Feller AC (1990) Histopathologie der Non-Hodgkin Lymphome, (nach der aktualisierten Kiel-Klassifikation). Springer, Berlin Heidelberg New York
Lennert K, Mohri N, Stein H, Kaiserling E (1975) The histopathology of malignant lymphoma. Br J Haematol 31 [Suppl]: 193–203
Lennert K, Mohri N, Stein H, Kaiserling E, Müller-Hermelink HK (1978) Malignant lymphomas other than Hodgkin's immunology. In: Uehlinger E (ed) Handbuch der speziellen pathologischen Anatomie und Histologie, Vol I/3B. Springer, Berlin Heidelberg New York
Lennert K, Knecht H, Burkkert M (1979) Vorstadien maligner Lymphome: Prelymphomas. Verh Dtsch Ges Pathol 63: 170–196
Leung CY, Ho FCS, Srivastava G, Loke SL, Liu YT, Chan ACL (1993) Usefulness of follicular dendritic cell pattern in classification of peripheral T-cell lymphomas. Histopathology 23: 433–437
Lindhout E, Mevissen MLCM, Kwekkeboom J, Tager JM, De Groot C (1993) Direct evidence that human follicular dendritic cells (FDC) rescue germinal center B cells from death by apoptosis. Clin Exp Immunol 91: 330–336
Liu Y-J, Joshua DE, Williams GT, Smith CA, Gordon J, MacLennan ICM (1989) Mechanism of antigen-driven selection in germinal centres. Nature 342: 929–931
Lukes RJ, Butter JJ (1966) The pathology and nomenclature of Hodgkin's disease. Cancer Res 26: 1063
Maeda K, Matsuda M, Narabayashi M, Imai Y (1993a) Follicular dendritic cells (FDC) in B-cell lymphoma—their distribution, morphology, phenotypes and functions. Dendrit Cells 2: 13–20
Maeda K, Matsuda M, Narabayashi M, Nagashia R, Degawa N, Imai Y (1993b) Follicular dendritic cells in malignant lymphomas—distribution, phenotypes and ultrastructures. Adv Exp Med Biol 329: 399–404
Mori N, Oka K, Kojima M (1988) DRC antigen expression in B-cell lymphomas. Am J Clin Pathol 89: 488–492
Naiem M, Gerdes J, Abdulaziz Z, Stein H, Mason DY (1983) Production of monoclonal antibody reactive with human dendritic reticulum cells and its use in the immunohistological analysis of lymphoid tissue. J Clin Pathol 36: 167–175
Nossal GJV, Abbot J, Michell J, Lummus Z (1968) Antigens in immunity. XV. Ultrastructural features of antigen capture in primary and secondary lymphoid follicles. J Exp Med 127: 277–288
Pallesen G, Myhre-Jensen O (1987) Immunophenotypic analysis of neoplastic cells in follicular dendritic cell sarcoma. Leukemia 1: 549–557
Parsonnet J, Hansen S, Rodriguez L, Gelb AB, Warnke RA, Jellum E, Orentreich N, Vogelman JH, Friedman GD (1994) Helicobacter pylori infection and gastric lymphoma. N Engl J Med 330: 1267–1271
Parwaresch MR, Radzun HJ, Hansmann M-L, Peters K-P (1983) Monoclonal antibody Ki-M4 specifically recognizes human dendritic reticulum cells (follicular dendritic cells) and their possible precursor in blood. Blood 62: 585–590
Petrasch S, Schmitz J, Pérez-Alvarez CJ, Brittinger G (1989) Reactivity of mAb of the B-cell panel with isolated follicular dendritic cells. In: Knapp W, Dörken B, Gilks WR, Rieber EP, Schmidt RE, Stein H, van dem Borne AEGK (eds) Leukocyte-typing IV. White cell differentiation antigens. Oxford University Press, Oxford, pp 185–187
Petrasch S, Pérez-Alvarez CJ, Schmitz J, Kosco M (1990) Antigenic phenotyping of human follicular dendritic cells isolated from nonmalignant lymphatic tissue. Eur J Immunol 20: 1013–1018
Petrasch SG, Kosco MH, Pérez-Alvarex JC, Schmitz J, Brittinger G (1991) Proliferation of germinal center B lymphocytes in vitro by direct membrane contact with follicular dendritic cells. Immunobiology 183: 451–462
Petrasch S, Kosco M, Pérez-Alvarez C, Schmitz J, Brittinger G (1992a) Proliferation of non-Hodgkin lymphoma lymphocytes in vitro is dependent upon follicular dendritic cell interactions. Br J Haematol 80: 21–26
Petrasch S, Kosco M, Schmitz J, Wacker HH, Brittinger G (1992b) Follicular dendritic cells in non-Hodgkin lymphoma express adhesion molecules complementary to ligands on neoplastic B-cells. Br J Haematol 82: 695–700
Petrasch S, Brittinger G, Wacker HH, Schmitz J, Kosco-Vilbois M (1994) Follicular dendritic cells in non-Hodgkin's lymphomas. Leuk Lymph 15: 33–43
Rademakers LHPM (1992) Dark and light zones of germinal centres of the human tonsil: an ultrastructural study with emphasis on heterogeneity of follicular dendritic cells. Cell Tissue Res 269: 359–368
Rappaport H, Winter WJ, Hicks EB (1956) Follicular lymphoma. A reevaluation of its position in the scheme of malignant lymphoma, based on a survey of 253 cases. Cancer 9: 792–821

Ratech H, Sheibani K, Nathwani BN, Rappaport H (1988) Immunoarchitecture of the "pseudofollicles" of well-differentiated (small) lymphocytic lymphoma: a comparison with true follicles. Hum Pathol 19: 89–94

Schnizlein CT, Szakal AK, Tew JG (1984) Follicular dendritic cells in the regulation and maintenance of immune responses. Immunobiology 168: 391–402

Schriever F, Freedman AS, Freeman G, Messner E, Lee G, Daley J, Nadler LM (1989) Isolated human follicular dendritic cells display a unique antigenic phenotype. J Exp Med 169: 2043–2058

Schwartz BA, Pinkus G, Bacus S, Toder M, Weinberg DS (1989) Cell proliferation in non-Hodgkin's lymphomas. Digital image analysis of Ki-67 antibody staining. Am J Pathol 134: 327–336

Sellheyer K, Schwarting R, Stein H (1989) Isolation and antigenic profile of follicular dendritic cells. Clin Exp Immunol 78(3): 431–436

Spencer J, Finn T, Pulford KAF, Mason DY, Isaacson PG (1985) The human gut contains a novel population of B lymphocytes which resemble marginal zone cells. Clin Exp Immunol 62: 607

Stauder R, Greil R, Schultz TF, Thaler J, Gattringer C, Radaskiewicz T, Dierich MP, Huber H (1989) Expression of leucocyte function-associated antigen-1 and 7F7-antigen, an adhesion molecule related to intercellular adhesion molecule-1 (ICAM-1) in non-Hodgkin lymphomas and leukaemias: possible influence on growth pattern and leucaemic behaviour. Clin Exp Immunol 77: 234–238

Stein H, Bonk A, Tolksdorf G, Lennert K, Rodt H, Gerdes J (1980) Immunohistologic analysis of the organization of normal lymphoid tissue and non-Hodgkin's lymphomas. J Histochem Cytochem 28: 746–760

Stein H, Gerdes J, Mason DY (1982a) The normal and malignant germinal centre. Clin Haematol 11: 531–559

Stein H, Uchánska-Ziegler B, Gerdes J, Ziegler A, Wernet P (1982b) Hodgkin and Sternberg-Reed cells contain antigens specific to late cells of granulopoiesis. Int J Cancer 29: 283–290

Stein H, Lennert K, Feller AC, Mason DY (1984) Immunohistological analysis of human lymphoma: Correlation of histological and immunological categories. Adv Cancer Res 42: 67–147

Tabrizichi H, Hansmann M-L, Parwaresch MR, Lennert K (1990) Distribution pattern of follicular dentritic cells in low grade B-cell lymphomas of the gastrointestinal tract immunostained by Ki-FDC1P: a new paraffin resistant monoclonal antibody. Mod Pathol 3: 470–478

Tiemann M, Parwaresch MR (1991) Diagnostischer Wert des neuen monoklonalen Antikörpers Ki-M4p in der Differenzierung maligner Lymphome. Verh Dtsch Ges Pathol 75: 303

Vaux DL, Cory S, Adams JM (1988) Bcl-2 gene promotes hemopoietic cell survival and cooperates with c-myc to immortalize pre-B cells. Nature 335: 440–442

Wacker HH, Radzun HJ, Mielke V, Parwaresch MR (1987) Selective recognition of rat follicular dendritic cells (dendritic recticulum cells) by a new monoclonal antibody Ki-M4R in vitro and in vivo. J Leukocyte Biol 41: 70–77

Weisenburger DD, Kim H, Rappaport H (1982) Mantle zone lymphoma: a follicular variant of intermediate lymphocytic lymphoma. Cancer 49: 1429–1438

Wood GS, Turner RR, Shiurba RA, Eng L, Warnke RA (1985) Human dendritic cells and macrophages. In situ immunophenotypic definition of subsets that exhibit specific morphologic and microenvironmental characteristics. Am J Pathol 119: 73–82

Subject Index

Current Topics in Microbiology and Immunology

Volumes published since 1989 (and still available)

Vol. 158: **Muzyczka, Nicholas (Ed.):** Viral Expression Vectors. 1992. 20 figs. IX, 176 pp. ISBN 3-540-52431-2

Vol. 159: **Gray, David; Sprent, Jonathan (Ed.):** Immunological Memory. 1990. 38 figs. XII, 156 pp. ISBN 3-540-51921-1

Vol. 160: **Oldstone, Michael B. A.; Koprowski, Hilary (Eds.):** Retrovirus Infections of the Nervous System. 1990. 16 figs. XII, 176 pp. ISBN 3-540-51939-4

Vol. 161: **Racaniello, Vincent R. (Ed.):** Picornaviruses. 1990. 12 figs. X, 194 pp. ISBN 3-540-52429-0

Vol. 162: **Roy, Polly; Gorman, Barry M. (Eds.):** Bluetongue Viruses. 1990. 37 figs. X, 200 pp. ISBN 3-540-51922-X

Vol. 163: **Turner, Peter C.; Moyer, Richard W. (Eds.):** Poxviruses. 1990. 23 figs. X, 210 pp. ISBN 3-540-52430-4

Vol. 164: **Bækkeskov, Steinnun; Hansen, Bruno (Eds.):** Human Diabetes. 1990. 9 figs. X, 198 pp. ISBN 3-540-52652-8

Vol. 165: **Bothwell, Mark (Ed.):** Neuronal Growth Factors. 1991. 14 figs. IX, 173 pp. ISBN 3-540-52654-4

Vol. 166: **Potter, Michael; Melchers, Fritz (Eds.):** Mechanisms in B-Cell Neoplasia. 1990. 143 figs. XIX, 380 pp. ISBN 3-540-52886-5

Vol. 167: **Kaufmann, Stefan H. E. (Ed.):** Heat Shock Proteins and Immune Response. 1991. 18 figs. IX, 214 pp. ISBN 3-540-52857-1

Vol. 168: **Mason, William S.; Seeger, Christoph (Eds.):** Hepadnaviruses. Molecular Biology and Pathogenesis. 1991. 21 figs. X, 206 pp. ISBN 3-540-53060-6

Vol. 169: **Kolakofsky, Daniel (Ed.):** Bunyaviridae. 1991. 34 figs. X, 256 pp. ISBN 3-540-53061-4

Vol. 170: **Compans, Richard W. (Ed.):** Protein Traffic in Eukaryotic Cells. Selected Reviews. 1991. 14 figs. X, 186 pp. ISBN 3-540-53631-0

Vol. 171: **Kung, Hsing-Jien; Vogt, Peter K. (Eds.):** Retroviral Insertion and Oncogene Activation. 1991. 18 figs. X, 179 pp. ISBN 3-540-53857-7

Vol. 172: **Chesebro, Bruce W. (Ed.):** Transmissible Spongiform Encephalopathies. 1991. 48 figs. X, 288 pp. ISBN 3-540-53883-6

Vol. 173: **Pfeffer, Klaus; Heeg, Klaus; Wagner, Hermann; Riethmüller, Gert (Eds.):** Function and Specificity of γ/δ TCells. 1991. 41 figs. XII, 296 pp. ISBN 3-540-53781-3

Vol. 174: **Fleischer, Bernhard; Sjögren, Hans Olov (Eds.):** Superantigens. 1991. 13 figs. IX, 137 pp. ISBN 3-540-54205-1

Vol. 175: **Aktories, Klaus (Ed.):** ADP-Ribosylating Toxins. 1992. 23 figs. IX, 148 pp. ISBN 3-540-54598-0

Vol. 176: **Holland, John J. (Ed.):** Genetic Diversity of RNA Viruses. 1992. 34 figs. IX, 226 pp. ISBN 3-540-54652-9

Vol. 177: **Müller-Sieburg, Christa; Torok-Storb, Beverly; Visser, Jan; Storb, Rainer (Eds.):** Hematopoietic Stem Cells. 1992. 18 figs. XIII, 143 pp. ISBN 3-540-54531-X

Vol. 178: **Parker, Charles J. (Ed.):** Membrane Defenses Against Attack by Complement and Perforins. 1992. 26 figs. VIII, 188 pp. ISBN 3-540-54653-7

Vol. 179: **Rouse, Barry T. (Ed.):** Herpes Simplex Virus. 1992. 9 figs. X, 180 pp. ISBN 3-540-55066-6

Vol. 180: **Sansonetti, P. J. (Ed.):** Pathogenesis of Shigellosis. 1992. 15 figs. X, 143 pp. ISBN 3-540-55058-5

Vol. 181: **Russell, Stephen W.; Gordon, Siamon (Eds.):** Macrophage Biology and Activation. 1992. 42 figs. IX, 299 pp. ISBN 3-540-55293-6

Vol. 182: **Potter, Michael; Melchers, Fritz (Eds.):** Mechanisms in B-Cell Neoplasia. 1992. 188 figs. XX, 499 pp. ISBN 3-540-55658-3

Vol. 183: **Dimmock, Nigel J.:** Neutralization of Animal Viruses. 1993. 10 figs. VII, 149 pp. ISBN 3-540-56030-0

Vol. 184: **Dunon, Dominique; Mackay, Charles R.; Imhof, Beat A. (Eds.):** Adhesion in Leukocyte Homing and Differentiation. 1993. 37 figs. IX, 260 pp. ISBN 3-540-56756-9

Vol. 185: **Ramig, Robert F. (Ed.):** Rotaviruses. 1994. 37 figs. X, 380 pp. ISBN 3-540-56761-5

Vol. 186: **zur Hausen, Harald (Ed.):** Human Pathogenic Papillomaviruses. 1994. 37 figs. XIII, 274 pp. ISBN 3-540-57193-0

Vol. 187: **Rupprecht, Charles E.; Dietzschold, Bernhard; Koprowski, Hilary (Eds.):** Lyssaviruses. 1994. 50 figs. IX, 352 pp. ISBN 3-540-57194-9

Vol. 188: **Letvin, Norman L.; Desrosiers, Ronald C. (Eds.):** Simian Immunodeficiency Virus. 1994. 37 figs. X, 240 pp. ISBN 3-540-57274-0

Vol. 189: **Oldstone, Michael B. A. (Ed.):** Cytotoxic T-Lymphocytes in Human Viral and Malaria Infections. 1994. 37 figs. IX, 210 pp. ISBN 3-540-57259-7

Vol. 190: **Koprowski, Hilary; Lipkin, W. Ian (Eds.):** Borna Disease. 1995. 33 figs. IX, 134 pp. ISBN 3-540-57388-7

Vol. 191: **ter Meulen, Volker; Billeter, Martin A. (Eds.):** Measles Virus. 1995. 23 figs. IX, 196 pp. ISBN 3-540-57389-5

Vol. 192: **Dangl, Jeffrey L. (Ed.):** Bacterial Pathogenesis of Plants and Animals. 1994. 41 figs. IX, 343 pp. ISBN 3-540-57391-7

Vol. 193: **Chen, Irvin S. Y.; Koprowski, Hilary; Srinivasan, Alagarsamy; Vogt, Peter K. (Eds.):** Transacting Functions of Human Retroviruses. 1995. 49 figs. IX, 240 pp. ISBN 3-540-57901-X

Vol. 194: **Potter, Michael; Melchers, Fritz (Eds.):** Mechanisms in B-cell Neoplasia. 1995. 152 figs. XXV, 458 pp. ISBN 3-540-58447-1

Vol. 195: **Montecucco, Cesare (Ed.):** Clostridial Neurotoxins. 1995. 28 figs. XI, 278 pp. ISBN 3-540-58452-8

Vol. 196: **Koprowski, Hilary; Maeda, Hiroshi (Eds.):** The Role of Nitric Oxide in Physiology and Pathophysiology. 1995. 21 figs. IX, 90 pp. ISBN 3-540-58214-2

Vol. 197: **Meyer, Peter (Ed.):** Gene Silencing in Higher Plants and Related Phenomena in Other Eukaryotes. 1995. 17 figs. IX, 232 pp. ISBN 3-540-58236-3

Vol. 198: **Griffiths, Gillian M.; Tschopp, Jürg (Eds.):** Pathways for Cytolysis. 1995. 45 figs. IX, 224 pp. ISBN 3-540-58725-X

Vol. 199/I: **Doerfler, Walter; Böhm, Petra (Eds.):** The Molecular Repertoire of Adenoviruses I. 1995. 51 figs. XIII, 280 pp. ISBN 3-540-58828-0

Vol. 199/II: **Doerfler, Walter; Böhm, Petra (Eds.):** The Molecular Repertoire of Adenoviruses II. 1995. 36 figs. XIII, 278 pp. ISBN 3-540-58829-9

Vol. 199/III: **Doerfler, Walter; Böhm, Petra (Eds.):** The Molecular Repertoire of Adenoviruses III. 1995. 50 figs. Approx. XIII, 324 pp. ISBN 3-540-58987-2

Vol. 200: **Kroemer, Guido; Martinez-A., Carlos (Eds.):** Apoptosis in Immunology. 1995. 14 figs. XI, 242 pp. ISBN 3-540-58756-X

Springer-Verlag and the Environment

We at Springer-Verlag firmly believe that an international science publisher has a special obligation to the environment, and our corporate policies consistently reflect this conviction.

We also expect our business partners – paper mills, printers, packaging manufacturers, etc. – to commit themselves to using environmentally friendly materials and production processes.

The paper in this book is made from low- or no-chlorine pulp and is acid free, in conformance with international standards for paper permanency.

MIX
Papier aus verantwortungsvollen Quellen
Paper from responsible sources
FSC® C105338

If you have any concerns about our products,
you can contact us on
ProductSafety@springernature.com

In case Publisher is established outside the EU,
the EU authorized representative is:
Springer Nature Customer Service Center GmbH
Europaplatz 3, 69115 Heidelberg, Germany

Printed by Libri Plureos GmbH
in Hamburg, Germany